Differentiating

in Algebra

preK–grade 2

Differentiating
in Algebra

A **Content Companion** for
- Ongoing Assessment
- Grouping Students
- Targeting Instruction
- Adjusting Levels of Cognitive Demand

Jennifer Taylor-Cox

HEINEMANN
Portsmouth, NH

Heinemann
361 Hanover Street
Portsmouth, NH 03801–3912
www.heinemann.com

Offices and agents throughout the world

Library of Congress Cataloging-in-Publication Data
Taylor-Cox, Jennifer.
    Differentiating in algebra, preK–grade 2 : a content companion for ongoing assessment, grouping students, targeting instruction, and adjusting levels of cognitive demand / Jennifer Taylor-Cox.
        p. cm. — (Differentiating in number & operations and the other math content standards, preK–grade 2)
    Includes bibliographical references and index.
    ISBN-13: 978-0-325-02184-3
    ISBN-10: 0-325-02184-8
    1. Algebra—Study and teaching (Early childhood)—Activity programs—United States.   2. Algebra—Study and teaching (Elementary)—Activity programs—United States. 3. Individualized instruction—United States.   4. Differentiation (Cognition)—United States.   5. Mathematical readiness.   6. Classroom management.   I. Title.   II. Title: Differentiating in algebra, preK–grade two.
    QA159.T39 2008
    372.7—dc22   2008017185

Editor: Emily Michie Birch
Production: Lynne Costa
Cover design: Jenny Jensen Greenleaf
Typesetter: Publishers' Design and Production Services, Inc.
Manufacturing: Louise Richardson

Printed in the United States of America on acid-free paper

12   11   10   09   08   VP   1   2   3   4   5

CONTENTS

CHAPTER 1

How to Use This Book

The Purpose of the *Algebra Content Companion*

Why is there a Content Companion?

The purpose of this *Content Companion* is to provide prekindergarten through second grade mathematics educators with the tools needed to target instruction through differentiation in algebra. The main book (*Number & Operations*) offers ideas, techniques, and strategies to use when targeting instruction in number and operations. The *Algebra Content Companion* allows educators to understand how to use these same ideas, techniques, and strategies when teaching algebra. The *Algebra Content Companion* offers specific content examination, student work samples, and lesson ideas. The rationale behind the content companions is that while number and operations is a critically important math standard, it is not the only standard that prekindergarten through second grade math educators teach. Algebra, geometry, measurement, and data and probability are also very important content standards.

How to Use This *Algebra Content Companion*

How do I use this book?

The easiest way to use this book is to follow the number and operations book, looking for the cross-references boxes in the margin that indicate where other mathematics content explanations and examples are offered in the *Algebra Content Companion*. Each time the reader sees this cross-reference box in the number and operations book, the reader can turn to the corresponding page in this content companion to find specific information pertaining to algebra. There are far too many books out there that attempt to explain to readers that similar strategies can be applied in all areas of mathematics, but these resources often fail to provide specific examples. Likewise, some resources attempt to cram all content into one resource, but these resources are often limited and incomplete. To address these problems, the *Content Companions* provide a comprehensive set of resources that prekindergarten through second-grade educators can use to target instruction through differentiation in mathematics.

The chapters in the *Algebra Content Companion* align with the chapters in the number and operations book. Chapter 1 offers an overview. Chapter 2 provides specific content information and examination pertaining to algebra. Chapter 3 directly addresses targeted instruction using informal preassessment. Chapter 4 involves grouping students. Chapter 5 tackles levels of cognitive demand. Chapter 6 focuses on learning frameworks. Chapter 7 provides personal assessment examples and situations.

While any one of the *Content Companions* could be used as a stand-alone resource, it is best understood within the context of the number and operations book. Educators are encouraged to use the number and operations book along with the four *Content Companions*—algebra, geometry, measurement, and data analysis and probability—to learn how to target instruction through differentiation in mathematics for prekindergarten through second-grade students.

CHAPTER 2

Algebra in Prekindergarten Through Second Grade

The Significance of Algebra

Algebra. Algebra? Algebra!

Algebra—many early childhood teachers cringe at the word. Others smile with fond memories. Because of or despite our emotional ties to algebra, we need to begin placing greater emphasis on algebra in the early years. Why? Mainly because algebra serves as a gate and a barrier for students later in their academic lives (Lott 2000). The gate to higher education is opened for students who successfully take algebra in middle school or in high school because colleges and universities require algebra. For others the gate is closed, creating a barrier for students who do not successfully take algebra. Moreover, notes Robert Moses, "People who don't understand algebra today are like those people who couldn't read or write in the industrial age" (Checkley 2001). Think of the equity issues involved. Doesn't every child deserve to be literate in mathematics? Doesn't every student deserve equal opportunity to higher education? The resounding answer is yes—algebra for all! However, "algebra for all" can become an empty promise without a plan of action. In order to prepare students to be successful in secondary-level algebra classes, we need to build the foundations of algebraic thinking from prekindergarten through second grade. The "big ideas" of algebra include understanding patterns, relations, and functions; representing and analyzing math situations; using models to represent quantitative relationships; and analyzing quantitative and qualitative change in various contexts (National Council of Teachers of Mathematics 2000). All of these aspects of algebra "enhance children's natural interest in mathematics and their disposition to use it to make sense of their physical and social worlds" (National Association for the Education of Young Children and National Council of Teachers of Mathematics 2002, 5).

Understanding Patterns, Relations, and Functions

Patterns provide numerous opportunities for algebraic thinking. Many early childhood teachers are experts at helping young children experience patterns. We highlight patterns on the calendar. We lead *clap-tap* pattern games. We point out patterns in fabrics and art. Patterns are a part of everyday life and we do a pretty good job at helping young children recognize and extend simple patterns. We ask, "Where's the pattern?" and "What comes next?" However, because the goals include encouraging students to "recognize, describe, extend, and translate patterns" (National Council of Teachers of Mathematics 2000, 90) we must not stop at recognizing and extending patterns. Describing patterns and translating patterns take more thought. Students need the algebraic reasoning and math vocabulary to describe a pattern. Students need even more reasoning and thought to translate a pattern, which involves recognizing that two different patterns have the same features—i.e., they are the same type of pattern. For example, " ◆ ☺☺ ◆ ☺☺ ◆ ☺☺" can translate into "● ▲ ▲ ● ▲ ▲ ● ▲ ▲." It is the same pattern made with different symbols. This is the very foundation of algebraic relationships. Understanding how patterns are related enables students to make predictions about patterns revealing pattern functions and generalizations.

Types of Patterns

There are several different types of patterns. The most common type of patterns that teachers of prekindergarten through second grade tend to focus on is the repeating pattern. In concordance with its name, a repeating pattern has a repetitive nature. As one of my students explained, "A repeating pattern goes over and over and over, again." The core of the pattern is present and then repeated multiple times in uninterrupted order. When we present patterns to students, we need to make sure the pattern core is repeated at least twice to show the nature of the pattern.

Many years ago my daughter was working on her first-grade math homework. Of course, I asked to see the assignment. Three shapes were given (■ ● ▼) and the directions were to "continue the pattern." The problem was that there was not a pattern to continue because the terms did not repeat. My daughter and I discussed how she could create a pattern using the three shapes, but that a repeating pattern did not yet exist. Using this information, my daughter decided to create the following repeating pattern, "■ ■ ■ ● ● ● ▼ ▼ ▼ ■ ■ ■ ● ● ● ▼ ▼ ▼ ■ ■ ■ ● ● ● ▼ ▼ ▼." She took the task to a higher level and all was right in the world—until she came home the next day with her paper marked with a big red X. Apparently the teacher wanted her to "continue the pattern" in this way: "■ ● ▼ ■ ● ▼ ■ ● ▼." There were two problems. First, the teacher's understanding of repeating patterns was minimal (although this lack of understanding was most likely generated by the publishers of the workbook that she copied the worksheet from). Second, my six-year-old daughter no longer believed that I was a good math teacher. Neither problem lasted long, but the experience prompted me to look deeper into the topic of early algebraic thinking—particularly patterns.

A variety of properties can be used to make repeating patterns. Patterns can be made using color, shape, size, texture, sound, and many other characteristics. Basically any descriptor that can be attached to an object can serve as a way to name that part of the repeating pattern. Playing pattern games is an excellent way to help young children learn more about repeating patterns. Many teachers have students engage in pattern games during transition times.

While Ms. Snyder's class comes to the front of the room for group time, they can participate in chants that include sound patterns. They use voice pitch and volume to chant, "Loud voice, soft voice, loud voice, soft voice, loud voice, soft voice . . ." Other times, Ms. Snyder uses nonsense words, "Zat, Tat, Tat, Zat, Tat, Tat, Zat, Tat, Tat . . ." or spells the next activity in chant form, "m, a, t, h, m, a, t, h, m, a, t, h . . ." There are so many transitions during the school day, why not use the time to gain further experiences with repeating patterns?

To be able to recognize and extend repeating patterns, children need to be able to sort objects. Knowing that the same color or the same size belongs in the same group is a foundational concept of patterning. Children will most likely have a difficult time working with patterns if they do not know how to sort the objects into groups that make sense. Sorting can be a complex concept for some children. When we give them a set of objects to sort, some children make groups of objects that do not have anything in common. I usually ask, "What do you call this group?" or "What's the name of the group you made?" By requiring the students to label the groups, teachers can help students understand what it means to sort.

> While participating in a seashell sorting activity I noticed a prekindergartner, Jenna, had two apparently random groups. I asked her to tell me the name of one of her groups. She replied, "It's the everything group." So I prompted her thinking further by asking what the other group was called. She responded, "Well, that's the everything group, too." I asked Jenna what was different about the two everything groups. Jenna thought for a moment and then replied, "This is everything I like and this is everything I don't like." What appeared as random groups to me was actually a grouping by Jenna's preferences. Through further dialogue and reflection, Jenna was able to describe what she liked about certain shells and what she did not like about certain shells. As it turned out, Jenna was fond of shells that were shiny inside, so she labeled her groups "shiny inside" and "not shiny inside." Encouraging children to label groups when sorting allows them to better understand the concept of sorting—which will, in turn, help them become better at recognizing, describing, extending, and translating repeating patterns.

Young children tend to focus on color as the dominant attribute. They create patterns such as "red, yellow, red, yellow, red, yellow . . ." Even when teachers attempt to demonstrate shape patterns, students may actually be focused on the colors. For example, the teacher uses pattern blocks to create a simple pattern—"hexagon, trapezoid, hexagon, trapezoid, hexagon, trapezoid . . ." When she asks one of the students to tell what comes next, the child exclaims, "Yellow!" Because all the hexagons in the set are yellow and all the trapezoids in the set are red, the child does not need to focus on shape to extend the pattern. However, if we use different-colored shapes to make the shapes in the pattern, we can help students focus on shape as the repeating attribute, as in Figure 2–1.

Repeating patterns are not the only type of patterns. There are also growing (ascending) and descending patterns. Growing patterns do not repeat. However, there is a repeating function within the growing patterns. Some constant change is happening to each term in a growing pattern. If we are working with numbers, the numbers increase by the same amount. If we are working with shapes, the shapes increase in size or quantity by some constant amount. Likewise, in a descending pattern the change is constant, but the size or mag-

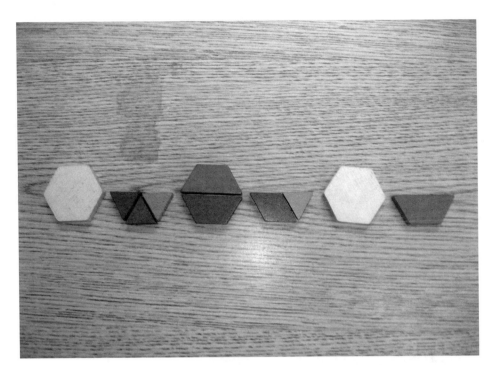

Figure 2–1 *Hexagon and Trapezoid Pattern*

nitude is decreasing. With very young children, growing and descending patterns can be exemplified with towers of blocks or snap cubes. Each tower is bigger (or smaller) than the preceding tower by the same number of blocks. For example, a tower of two, a tower of four, a tower of six, and so on is a growing pattern that increases by two (see Figure 2–2).

A kindergartener, Justice, made a descending pattern using bears and plates. He started with four bears on a plate, and then three bears on a plate, two bears on a plate, and one bear on a plate (see Figure 2–3). Justice announced, "This is a getting littler pattern!"

Growing and descending patterns can also be formed using numbers. When students count by tens, they are working with growing patterns. Skip counting is a common and productive activity used in many classrooms. I once asked a first grader to skip count by tens. He successfully completed the task from ten to one hundred. I asked him what came next and he replied, "Nothing. It's over." His response taught me to help students see that skip counting can go on forever. If we stop at one hundred every time, students can develop a misconception that the counting ends at a certain point. Likewise, we don't want students to think that you can only skip count starting at zero. How do you skip count by fives starting with two? How do you skip count by tens starting at 103? How do you skip count by tens backward starting with ninety-nine? What is the third term when you skip count by fives starting with four?

Figure 2–2 *A Growing Pattern Using Snap Cubes*

Figure 2–3 *A Student-Created Descending Pattern*

These questions and others help students gain a broader understanding of growing and descending patterns.

As students notice the relationships between and among the numbers in patterns, they begin to uncover the function that formulates the pattern. Sometimes young children understand the constant change in an ascending or descending pattern as a *secret*, something to be discovered through examining the relationships of the terms.

Numeric patterns can be categorized as arithmetic and geometric. In essence, arithmetic patterns have a common difference and geometric patterns have a common ratio. Arithmetic patterns have a constant function that involves addition or subtraction. The skip counting patterns in the previous paragraph fall under the arithmetic type of pattern. However, geometric patterns use multiplication or division. For example, "1, 2, 4, 8, 16, etc." is a pattern that does not involve adding. Instead each term in the sequence is multiplied by two to get the next term. A first grader who has not yet learned about multiplication may understand this sequence as *doubling each number*.

A couple of years ago, I was working with a group of students in second grade who needed some challenge in the area of growing/descending patterns. I posed this sequence to the group—"32, 16, 8, 4. . . ." The students used number lines and counters as they discussed the possible relationships of the numbers. They came up with the idea that the numbers were "being cut in exactly half." Even though they had not yet learned division, they were able to generalize about the pattern and decide that "2" must be the next number in the sequence.

Patterns can also be nonlinear. The size of the terms in a pattern may increase in a spiral fashion, as is present in some artwork, fabric, and tiling. The nautilus shell provides a mathematically brilliant concentric pattern represented in the size of the chambers within the spiral (the ratio is a perfect 1 to 1.618—the proportion is often called the *golden ratio*). Even if we are not teaching specific concentric patterns to our young students, it is still important for them to realize that not all patterns are linear.

Patterns, relations, and functions come alive in the early childhood classroom as students work with a variety of pattern types, discerning how the terms are related and what generalizations can be made about the pattern. As children learn to recognize, extend, describe, and translate simple patterns, they set the groundwork needed to move on to more complex algebraic thinking.

Representing and Analyzing Math Situations

Because representing and analyzing math situations and structures involve using math symbols and properties of mathematics such as associativity and commutativity, some educators may believe that this area of mathematics is not applicable to early childhood. To see how this part of algebraic thinking is actually quite germane to early mathematics, we need to investigate the foundational concepts.

Math Symbols

Math symbols help us represent and communicate math ideas. We need symbols to make the math more efficient. Instead of writing *six and two more will make a total of eight*, we can just write 6 + 2 = 8. Using the symbols makes communicating the math idea less cumbersome. The issue is that young children need to understand the concepts before moving to symbolic representation. They need to understand the meaning of addition and equality. If we move to symbols before these concepts are understood, children fall into robotically completing procedures instead of understanding the actual mathematics. For example, a first grader sees 4 + 3 = □ and thinks *Oh yeah, + means I count to four and count on three more—seven is the answer.* The student would arrive at the correct answer. However, without understanding the concept of addition, the procedure may be inappropriately applied to another situation such as □ + 3 = 7 where the child thinks *Oh yeah, + means I count to three and count on seven more—ten is the answer.* To accurately solve the latter situation, students need to understand addition, symbols, and equations. The procedure (prompted by the plus sign) of putting together the two given numbers does not work when the start of the equation is the unknown. While procedures are, no doubt, important in mathematics, memorizing procedures without the grounding in what those procedures actually mean may produce a hollow, artificial type of comprehension. This is why early childhood educators need to focus on concepts prior to procedures and symbols.

Using real objects in context helps young children build conceptual knowledge. If we are adding apples, we need to use real apples. If we are subtracting crayons, we need to use real crayons. If we are trying to find out how many students in the class are wearing shoes that tie, we need to look at everyone's shoes and decide how to group the people. Using real objects helps make abstract math concepts more concrete.

Years ago I was working in an elementary school with a few teachers who were being asked to decrease the focus on workbooks and pencil-paper tasks and increase hands-on, inquiry-based learning. We had introduced many new math manipulatives to the teachers and students. I visited the kindergarten teacher who was working on addition with a group of students. The problem they were working on was 2 + 3 = 5. The teacher took out nine ice cream sticks and demonstrated the math situation to the students in the way shown in Figure 2–4.

Yes, she used ice cream sticks to make the plus sign and the equal sign. And yes, this posed a huge problem for the students. There were nine ice cream sticks, yet the "answer" was five. They did not know what the symbols meant. They did not know which sticks to count and which sticks not to count. They stared at their teacher. Exasperated, she took out five more ice cream sticks and put them at the end of the absurd equation and announced, "This is what you do." After a minute or so, one student smiled and said, "Fourteen!"

Obviously, this is not the way to teach young children how to represent and analyze math situations and structures. The teacher's sole concentration was on procedure, without any regard to the mathematics represented by the manipulatives. She was simply writing the equation—symbols and all—with ice cream sticks. The symbols need to represent ideas, not confuse the math situation. Fortunately, the teacher did not give up. Afterward, we talked about the lesson. Eventually, she did come to understand how to accurately use manipulatives to represent math situations. She also learned to build understanding of the math concepts before introducing the symbols. The students learned to model math situations in ways that made sense. For example, the students placed the ice cream sticks on plates to show the two groups and discussed the process for adding two plates of ice cream sticks together (see Figure 2–5).

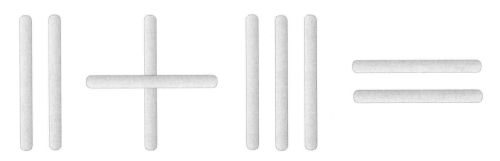

Figure 2–4 *Confusing Equation Formed with Ice Cream Sticks*

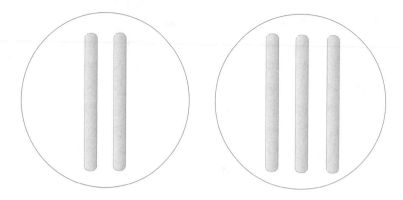

Figure 2–5 *A Better Use of Ice Cream Sticks to Model the Same Situation*

Math Properties

Even if we don't think we teach the commutative or associative properties to our young students, we do. Math operations and their properties are inseparable. Knowing that 6 + 4 gives you the same sum as 4 + 6 is the realization of the commutative property. Knowing that 5 + 3 + 5 is the same as 5 + 5 + 3, which is the same as 10 + 3, which equals 13 is the heart of the associative property. We teach these mental computation strategies to students to help them become more efficient with their basic facts, but we are also introducing corresponding mathematical properties. As noted by the National Council of Teachers of Mathematics (NCTM) (2000), "Students often discover and make generalizations about other properties. Although it is not necessary to introduce vocabulary such as commutativity or associativity, teachers must be aware of the algebraic properties used and understood by students at this age [prekindergarten through second grade]" (93).

A major concept embedded within mathematics properties is the art of composing and decomposing numbers. Composing numbers means building numbers and decomposing numbers means breaking numbers apart. Clements (Clements and Sarama 2004) explains that "composing and decomposing are combining and separating operations that allow children to build concepts of 'parts' and 'wholes.'" When children see a number as made up of specific related parts, they understand that number more completely.

The teacher asks a first grader, Marci, how to make the number 12. Marci reaches for a pencil and writes the number. Writing the number does not necessarily mean she truly understands the number, although we certainly know that she can communicate the number. Asking Marci to show and explain *how* the number 12 is made may provide more information about whether Marci actually understands 12. Let's say Marci shows 12 with 1 ten and 2 ones. The next step is to ask Marci if she knows any other ways to show 12. We want to see if she understands the many ways to compose 12. Marci may show 6 and 6 as a way to compose 12. The teacher may ask Marci this question, "If there are 12 counters and two fall on the floor, how many will be left on the table?" This question allows the teacher to see how fluent Marci is with decomposing the number 12.

Children can compose and deompose numbers when they are computing. If Marci sees 12 + 3, she may break apart 12 into 10 + 2 and then add 3 to obtain the sum. In this way Marci is decomposing numbers and using the associative property as she analyzes the mathematics situation and structure.

Equality

I often ask students, "What does *equal* mean?" The answers they give are very interesting. Some say, "*Equal* means the answer will be next." Some children say, "*Equal* means two lines." These explanations show a lack of understanding of equality. As noted by many researchers (Behr, Erlwanger, and Nichols 1980; Falkner, Levi, and Carpenter 1999; Senz-Ludlow and Walgamuth 1998), "children often have serious misconceptions about the meaning of the equals sign. Children tend to perceive the equals sign as a stimulus for an answer" (Molina and Ambrose 2006, 111). Without meaning, the equals sign is useless.

"Equality is an important algebraic concept that students must encounter and begin to understand in the lower grades" (National Council of Teachers of Mathematics 2000, 94). Equality involves "recognizing, defining, creating, and maintaining balance" (Taylor-Cox 2003). Using pan balances allows us to approach the teaching of equality in a hands-on fashion. Even very young children can begin to understand that when one side of the balance goes down, it contains more mass. Likewise, when the balance is horizontal, the pans hold equal mass.

Having worked with real pan balances for about a week, the prekindergartners in Mr. Rob's class knew how to recognize equality and balance. Mr. Rob demonstrated, pretending to be a pan balance, as he shared, "This hand holds a heavy rock and this hand holds a feather. Show me how the balance looks." Nearly all of the students leaned to one side (some even grunted at the pretend weight of the rock) to show that the rock weighed more. "Now this time we have an apple in each hand and the apples weigh exactly the same. Show me what the balance looks like," announced Mr. Rob. Nearly all of the students immediately stood with arms straight out. The few who were off-balance quickly joined the ranks. "We are balanced! We are balanced! We are balanced!" chanted the class. Mr. Rob's class has solid understanding of balance, which will help them form an understanding of equality.

One particular group of students in Mr. Rob's class was ready for an extra challenge. During center time, Mr. Rob worked with the small group. He asked one of the students (Timmy) to stand and pretend to be a balance. He placed three cubes in Timmy's right hand and asked the rest of the group how many cubes he should place in Timmy's left hand if he wanted Timmy to be balanced. Everyone in the group said, "Three." Then Mr. Rob posed this situation: "If I take two cubes out of Timmy's left hand, will he be balanced?" The group responded, "No." "How can we make Timmy balanced?" Mr. Rob asked. The group thought and one child spoke up, "Take two out of the other hand." Another child added, "Or you just put those back." The other children agreed. In this simple exercise, these prekindergarten students were engaging in recognizing, defining, creating, and maintaining balance. Algebraic thinking in action!

Roughly translated, *equal* means the "same as." What we want students to understand is that equations are set up, by design, to display that one side is the "same as" or the "same value as" the other side. Sometimes students are perplexed when they look at a problem such as $\square = 2 + 3$. One first grader looked at this problem and started laughing. Amid his snickers, he explained, "That's not right. You can't write it that way because the box has to go at the end." This particular student who was new to our school at the time was confident with addition, but only when the unknown was at the end (the result) of the equation. His understanding of equality needed to be developed.

Other times when students are faced with solving equations such as $2 + 5 = 3 + \square$, they add $2 + 5 + 3$ and fill in the blank with 10. When this happens, the teacher may say the child wasn't paying attention or the child did the problem too quickly. But it may actually be a lack of understanding of equality or the symbol for equal that is behind the error. To help develop meaning of equality, we should use the terms *equal* and the *same as* interchangeably. We want the students to become comfortable with the concept of equality. We also need to build an understanding of what is not equal or not the same as. Understanding what is not equal actually helps fortify the understanding of what is equal. As one of my former kindergartners said, "If it's not sane, it's not eagles!" *Good thing eagles are sane.*

Use Models to Represent Quantitative Relationships

In prekindergarten through second grade focusing on using models to represent quantitative relationships is not difficult. Early childhood educators typically know the importance of hands-on experiences, so they tend to use a lot of models in mathematics. However, we can easily fall into a trap of using only one model to represent a particular quantitative relationship.

> Ms. Newman worked with a group of second-grade students who were struggling with this algebraic situation.
>
> *Motorcycles have 2 wheels and cars have 4 wheels. If there are 16 wheels, how many cars and how many motorcycles are there?*
>
> Ms. Newman spoke as she demonstrated on the hundred chart. "Say the first one is a car. We should cross out 1, 2, 3, 4 wheels. The next one could be a motorcycle, so we should cross off how many wheels?" The group said "2" in unison. Ms. Newman smiled and asked, "What should we do next?" No one responded. Ms. Newman called on Anthony. "What should we do next, Anthony?" Anthony was quiet and then replied, "We should get some chips and use them as wheels to figure it out."

In this situation, the model Ms. Newman used did not connect with how the students, or at least Anthony, wanted to represent the quantitative relationships in the story problem. He wanted to make groups of 4 and groups of 2 out of chips to derive his answer. It doesn't mean that the hundred chart was the wrong model. It just means that students need to have the freedom to use various models to represent and, ultimately, understand the math.

Using models to represent and understand quantitative relationships should be part of the real-life mathematical experiences of children.

> In a kindergarten class, the students discuss art project choices. Each student is allowed to pick five stickers to use in an art project. There are pumpkins, leaves, and cats. Mary chooses two pumpkins, two leaves, and one cat. Stephen chooses four pumpkins and one cat. James chooses five cats. The students engage in math dialogue as they compare the various ways to make five.
>
> *Mary:* I have five.
>
> *Stephen:* I have five.

James: I have five, too.

Mary: My five looks different because I have everything.

Stephen: My five looks bigger because I have a lot of pumpkins.

Mary: But it's still five.

James: Yep it is. Just like mine is still five with no pumpkins and
no leaves.

Stephen: I guess fives can be different, but still five.

The math talk these kindergartners engaged in was filled with comparisons of some of the ways five is composed. The students studied and discussed different ways to model the number 5, including 2 + 2 + 1, 4 + 0 + 1, and 0 + 0 + 5. These are the kinds of situations that encourage young children to use mathematical models to represent and understand quantitative relationships.

Analyzing Qualitative and Quantitative Change

Algebraic thinking also includes analyzing change in various contexts. "The understanding that most things change over time, that such changes can be described mathematically, and that changes are predictable helps lay a foundation for applying mathematics to other fields and for understanding the world" (National Council of Teachers of Mathematics 2000, 95).

Algebraic change can be described as quantitative or qualitative. The distinction is in the use of exact or approximate amounts. Quantitative change has specific values. For example, during a thunderstorm the water in the rain gauge increases by 0.2 inches every hour or the bean plant grows one centimeter each day. On the other hand, qualitative change is determined by less exactness. For example, the child's shoe size increases as they get older or the dog's weight increases as it changes from a puppy to an adult dog. These changes are marked by steady increase. Yet some forms of change occur as continual decreases. Descriptions of the amount of water in a pot diminishing as the water boils and evaporates are examples of qualitative decreasing change. If we measure specifically how much water evaporates at certain time intervals, we can understand the decreasing change as quantitative.

In a first-grade classroom, George's job of the week is sharpening pencils (fun job because of the electric pencil sharpener—but the rule is that each pencil can only go in the sharpener for three seconds). George notices that the pencils get smaller each time he sharpens them. This observation can be understood as qualitative change. The teacher encourages the algebraic thinking by asking George to follow the "life" of a pencil for a week. George records the length of the pencil at the end of each day in his math journal. "Monday the pencil is 18 cm long. Tuesday the pencil is 17 cm long. Wednesday the pencil is 16 cm long. Thursday the pencil is 15 cm long. Friday the pencil is 14 cm long." The teacher asks George to make a prediction about how long the pencil will be at the end of the day on the following Monday. George uses the data he collected to make a prediction. George reports to the class, "I think the pencil will be 13 cm long on Monday." The teacher asks, "Why?" George responds, "Because every day it gets smaller by 1 centimeter. You have to skip Saturday and Sunday because we're not here." George describes the situation mathematically and uses the data to generalize and make predictions. In this situation, the change was first described qualitatively— "the pencil gets smaller every day." Then the change was described quantitatively—"the pencil gets smaller by 1 centimeter each day."

Not all situations lend themselves to moving from qualitative change to quantitative change. Sometimes all you need is the qualitative explanation to engage in mathematical descriptions, generalizations, and predictions. Consider this classroom scenario. In a prekindergarten classroom, the students record the daily weather on the calendar. The choices are hot, warm, cool, and cold. One of the students, Alistair, shares that "it used to be warm, now it is cool, cool, cool, every time." The teacher asks Alistair to use the data to make a prediction: "What do you think the weather will be tomorrow?" Alistair says, "I think it will be cool, again." The teacher asks, "Why?" Alistair responds, "I don't know." Another student raises her hand and replies, "It's getting cooler because winter is coming. Soon it will be cold." These students are investigating qualitative change as they notice the temperature decreasing over time.

A group of second graders works with quantitative change with number patterns. The students engage in mathematics discourse as they come to understand the relationships within the pattern. Rich and Kayla discuss the following pattern:

. . . 42, 47, 52, 57, 62, 67, 72, 77 . . .

Rich: The numbers are getting bigger.

Kayla: Every two numbers have the same digit in the tens.

Rich: Yeah, and then it goes up to the next ten.

Kayla: I think 82 is next.

Rich: Me too. Then 87. Then 92.

Kayla: Then 97. But how do we explain the change?

Rich: Well, we know it is adding.

Kayla: If you look at every other number it is adding 10.

Rich: Oh, and look if you do half of 10 it is 5.

Kayla: Yes. I see it. The numbers are changing by 5 every time.

Rich: When does it stop?

Kayla: I'm not sure. I guess it never stops.

In this exchange of math ideas, the students are analyzing, predicting, and generalizing about the quantitative change that occurs within the arithmetic pattern.

The four "big ideas" of algebra are quite connected. Analyzing change is linked to understanding patterns. Representing and analyzing math situations is intertwined with using models to represent quantitative relationships. All four ideas have strong connections. As prekindergarten through second-grade students continue to think algebraically, the foundations are secured for higher levels of understanding.

As educators, we need to give ample time for our students to experience different types of patterns. Students also need to represent and analyze math situations in ways that help them understand math properties and symbols. The use of models to represent and understand quantitative relationships is imperative. Likewise, students need to explore and analyze change in a variety of contexts. By providing these algebraic experiences for our students, we will help open the gates and diminish the barriers to higher levels of understanding. Algebra for all will be a reality because we have established the necessary groundwork in algebraic thinking.

CHAPTER 3

Targeting Instruction Using Index Questions in Algebra

Index Question for Analyzing Change

What kind of change is going on?

Ms. Hall teaches prekindergarten. She has twenty children in her class. The children have varying levels of mathematics ability. Currently Ms. Hall is teaching the children how to analyze qualitative change as part of the unit on algebraic thinking. While the students were working in centers, Ms. Hall interviewed each student using an index question. She made the instructional groups based on the students' responses.

See:
Number, p. 53
Data, p. 32
Geometry, p. 32
Measurement, p. 29

Explanation of the Analyzing Change Index Question

The index question in Figure 3–1 is aimed at revealing an understanding of how to analyze change. When teaching this component of algebra to young students, we want them to learn that change is often predictable. Using the given information, what is most likely to come next? This concept is connected to growing patterns, measurement, and sequencing.

Student Responses

Nathalie's Response

Nathalie's response to the index question is fascinating (see Figure 3–2). While the answer she specified is not correct, her explanation shows a good

See:
Number, p. 53
Data, p. 34
Geometry, p. 34
Measurement, p. 30

The flowers are changing.

Which flower should be next? Why?

Figure 3–1 *Index Question for Analyzing Change*

The flowers are changing.

Which flower should be next? Why?

Teacher: The flowers are changing. Which flower should be next?

Nathalie: This one [points to the medium-sized flower].

Teacher: Why?

Nathalie: Because the flowers get big and then they start to die.

Figure 3–2 *Nathalie's Response*

understanding of potential change. Flowers do, eventually, wither and die—which represents real-life change. The important element in this index question is the answer to the question, "Why?" Without Nathalie's explanation, her response indicates a low understanding of change. Because of the explanation, Ms. Hall knows Nathalie's understanding of this concept is actually higher. There were four other students who chose the medium-sized flower; however, these students did not give reasons like Nathalie's reason. The students needed more practice with sequencing and order by comparison. Ms. Hall decided to group Nathalie with the group who needed challenge because of the understanding she showed in her explanation.

Gloria's Response

Gloria shows a high level of understanding in her response, shown in Figure 3–3. She knows that the change shown in the picture is an increase of size. It makes perfect sense to her that the next flower should be the biggest one. There were ten other students that gave responses similar to Gloria's response. Ms. Hall was a bit surprised that so many of her students could analyze change so well. She learned that she didn't need to spend the next several days teaching this concept to the whole class because most of the students understood the idea. She could raise the level of complexity and offer a more challenging task to these students (including Gloria and Nathalie). In this way, differentiated instruction actually saves time for the teacher and the students because we do not need to teach something that students already know.

John's Response

John's response in Figure 3–4 is amusing. He randomly decides that the next flower should be the smallest one because *he wants it to go there*. This kind of response is not uncommon from young children. Sometimes they focus on something they like and decide that it should be the correct answer. This is part of the egocentric nature of children. It does not mean that John is a struggling student, what it means is that John (and the three other students who chose the smallest flower for random reasons and the other students that chose the medium-sized flower for random reasons) need some targeted instruction from the teacher to uncover the gaps keeping them from understanding how to analyze change.

The index question Ms. Hall used provided her with valuable information. Based on the evidence, Ms. Hall adjusted the level of cognitive demand and targeted instruction for two groups of students. She knew that the students working at a high level needed challenge and the students who did not yet understand this concept needed additional support and scaffolding. Ms. Hall met

The flowers are changing.

Which flower should be next? Why?

Teacher: The flowers are changing. Which flower should be next?
Gloria: The really big one.
Teacher: Why?
Gloria: Because the flowers are getting bigger.

Figure 3–3 *Gloria's Response*

The flowers are changing.

Which flower should be next? Why?

Teacher: The flowers are changing. Which flower should be next?
John: The little one.
Teacher: Why?
John: I want it to go there.

Figure 3-4 *John's Response*

with each group and was able to gear her teaching strategies to meet each group's specific needs without spending too much time on the process.

Small Group, Targeted Instruction

The responses given on the preassessment that Ms. Hall presented to her prekindergartners indicated that many of the children understand how to analyze change and some of the children need more experience with analyzing change (including sequencing and ordering by comparison). Ms. Hall meets with both small groups separately.

See:
Number, p. 56
Data, p. 37
Geometry, p. 37
Measurement, p. 34

The First Group Works with Analyzing Change in Various Contexts

This group of students (including John) works with Ms. Hall on showing growing patterns with towers of different-colored blocks. Ms. Hall presents three towers (a tower of one, a tower of two, and a tower of three) to the students. She asks them to discuss what the next tower should look like. The children share a few ideas that focus on the color of the blocks. Ms. Hall modifies her growing pattern to use only blocks of the same color to neutralize the distraction. This adjustment is beneficial. The students are able to analyze the change in the growing pattern and explain why the next tower should be larger by one. Next, Ms. Hall presents beads and string. She asks the students to string the beads in a growing pattern. She demonstrates how the pattern could begin and invites the children to make and then share other bead patterns. Most of the children are able to construct accurate growing patterns with the beads. However, a couple of the students make designs rather than growing patterns. Ms. Hall knows that these students will need more experiences with growing patterns and analyzing change in various contexts.

The Second Group Works with Analyzing Change in Various Contexts

Ms. Hall encourages the first group of students to continue working on the bead patterns independently while she meets with the second group (including Gloria and Nathalie) to provide a challenge. Ms. Hall knows that these students already understand how to analyze simple change. She can compact the information her curriculum guide indicates should be taught and forward-map into a more intricate task for the students. Ms. Hall decides to *bump up* the level by teaching this group of students how to analyze change in descending patterns. She presents a block pattern that includes towers that decrease in height (tower of five, tower of four, tower of three) and asks the students to describe what the next tower of blocks should look like. Ms. Hall encourages the students to share their ideas and then invites them to construct their own descending patterns using snap cubes. Ms. Hall has pairs of students work together to construct these patterns. The pairs then take a *gallery walk* to analyze the changes present in each other's patterns.

In this differentiated mathematics lesson, Ms. Hall targets the instruction for both groups of students. She compacts, scaffolds, and uses back-mapping and forward-mapping. The lessons are tiered to meet the needs of the students. The tasks are similar, yet they address different levels of understanding.

Index Question for Understanding Quantitative Relationships

Understanding quantitative relationships

See:
Number, p. 61
Data, p. 40
Geometry, p. 41
Measurement, p. 38

Ms. Shelby's class of twenty-four kindergartners is beginning to study the component of algebra that involves using models to represent and understand quantitative relationships. The students have had many experiences with using models in mathematics. They nearly always use manipulatives or pictures to explain math thinking. So while the algebraic topic Ms. Shelby is starting with her students is new, the use of models in mathematics is familiar.

Explanation of the Quantitative Relationships Index Question

Ms. Shelby uses a simple index question (see Figure 3–5) that serves as a pre-assessment for the study of algebra. The algebraic situation given to the students involves representing and understanding how many tricycles and bicycles are needed to have a total of twelve wheels. Obviously the algebra problem has several different correct answers. What Ms. Shelby is looking for is how the students use models (manipulatives or pictures) to represent and understand their answers.

See:
Number, p. 61
Data, p. 41
Geometry, p. 41
Measurement, p. 41

Student Responses

Jasmine's Response

Jasmine's response in Figure 3–6 indicates that she understands how many wheels are on one bicycle and how many wheels are on one tricycle. Jasmine's

Name _____

Tricycles and Bicycles

Tricycles have 3 wheels. Bicycles have 2 wheels.

If there are 12 wheels, how many tricycles and how many bicycles are there?

Draw a picture to show your answer.

Figure 3–5 *Quantitative Relationships Index Question*

Index question: Understanding quantitative relationships (kindergarten)

Index Question   Name

Tricycles and Bicycles

Tricycles have three wheels. Bicycles have two wheels.

If there are 12 wheels, how many tricycles and how many bicycles are there?   2

Draw a picture to show your answer.

Figure 3–6 *Jasmine's Response*

picture shows three small wheels grouped together and two larger wheels grouped together. While her answer is incorrect, her use of models to represent her answer is sound. Jasmine, along with many other students who answered similarly, is ready for on-grade-level instruction in this area of algebraic thinking. Ms. Shelby grouped the students together and implemented a mini-lesson on using models to represent and understand quantitative relationships.

Amy's Response

At first glance, the response given by Amy in Figure 3–7 may seem inaccurate. But a closer look with an early childhood educator's lens reveals a high level of understanding coupled with precise representation. Amy has two groups of three small sketched circles and three groups of two larger sketched circles. Her answer, *E BKS*, stands for "3 bikes" and *S tsrk* stands for "2 trikes." Amy understands and represents the quantitative relationships in this algebra situation. Amy and the other students who showed similar levels of understanding need to be challenged.

Figure 3–7
Amy's Response

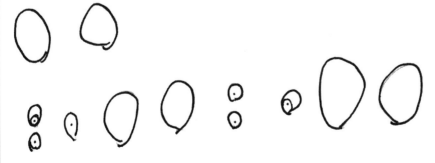

Index Question   Name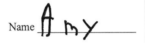

Tricycles and Bicycles

Tricycles have three wheels. Bicycles have two wheels.

If there are 12 wheels, how many tricycles and how many bicycles are there? E BKS

Draw a picture to show your answer. StShK

Sadriel's Response

Sadriel made an attempt to answer and represent the solution to the algebra problem (see Figure 3–8). He drew four wheels, yet his answer was three. Ms. Shelby needs to work with Sadriel and the other students who responded in a similar fashion. Using a small group structure and targeted instruction, Ms. Shelby plans to address the misconceptions held by these students.

The index question enabled Ms. Shelby to preassess her students' levels of knowledge. Most of the students needed instruction on how to accurately represent and understand quantitative relationships. Because all of the students drew some type of representation for the wheels, Ms. Shelby did not need to focus on initiation of models to solve math problems. For Sadriel's group, Ms. Shelby worked on how to make the model match the answer. For Jasmine's

Figure 3–8
Sadriel's Response

Index question: Understanding quantitative relationships (kindergarten)

Index Question   Name  Sadiel

Tricycles and Bicycles

Tricycles have three wheels. Bicycles have two wheels.

If there are 12 wheels, how many tricycles and how many bicycles are there?   3

Draw a picture to show your answer.

group, Ms. Shelby concentrated attention on how to use the models to solve the problem. For Amy's group, Ms. Shelby offered a challenge situation. All of the students engaged in activities that met their academic needs.

Small Group, Targeted Instruction

There were three levels revealed in the responses given on the preassessment Ms. Shelby gave her kindergartners. Ms. Shelby met with each group to target instruction.

See:
Number, p. 64
Data, p. 44
Geometry, p. 46
Measurement, p. 42

The First Group Works with Using Models to Represent Quantitative Relationships

This group of students (including Jasmine) appears to have some understanding of how to use models to represent an answer, but does not yet understand how to solve the problem accurately. With this group, Ms. Shelby uses the same problem that was presented on the preassessment. The students work with partners using twelve buttons (representing wheels) to make groups of two and groups of three on work mats. Afterward, Ms. Shelby has the students walk around and look at how each pair of students grouped the wheels. Using this *gallery walk* strategy allows the students to see and discuss similarities and differences in how various students represented the quantitative relationships in this situation.

The Second Group Works with Using Models to Represent Quantitative Relationships

Ms. Shelby focuses her attention on the next group. The students in this group (including Amy) need a challenge. They already know what Ms. Shelby's curriculum guide indicates that she should teach. Therefore Ms. Shelby needs to *bump up* the content for these particular students. One simple yet effective way to do this is to change the number of wheels from twelve to something greater. Ms. Shelby chooses eighteen because it is a multiple of three and a multiple of two. Additionally, Ms. Shelby instructs the students to find at least three different solutions (and representations) for the new math situation. In this way Ms. Shelby is forward-mapping to a higher level of content for these students.

The Third Group Works with Using Models to Represent Quantitative Relationships

Ms. Shelby worked with the third group (including Sadriel), back-mapping the content to a previous level to help the students establish stronger foundations in understanding quantitative relationships. Ms. Shelby needs to help the students understand how to represent one bike and one trike before moving to a higher level of complexity. She directs the students in a game involving a spinner that has a tricycle on one side and a bicycle on the other side. Ms. Shelby spins the spinner and the students show the number of wheels by placing buttons on their work mats. After a few spins, Ms. Shelby has the students continue to show the number of wheels—but now also encourages the students to *say* the number. After a few more rounds, Ms. Shelby asks the students to represent two bicycles and then two tricycles with the buttons. She encourages them to count the total number of wheels (buttons) and describe the situation—for example, "Two tricycles mean six wheels." Ms. Shelby continues to scaffold the instruction and encourage the students to talk about the number of bicycles, tricycles, and wheels.

In this differentiated mathematics lesson, Ms. Shelby successfully challenged and supported each of the groups through targeted instruction. Ms. Shelby had options for how to differentiate the instruction and chose to tier the lesson, use back-mapping, use forward-mapping, scaffold, compact, and activate prior knowledge. All of the students gained knowledge of how to use models to represent quantitative relationships.

Index Question for Repeating Patterns

Repeating Patterns Repeating Patterns Repeating Patterns

See:
Number, p. 69
Data, p. 49
Geometry, p. 52
Measurement, p. 48

Ms. Smith teaches a class of twenty-four first graders. The students range in ability and mathematics expertise. The next mathematics topic scheduled is algebra, specifically repeating patterns. Before beginning instruction, Ms. Smith developed a preassessment aimed at revealing the different levels of knowledge of the students.

Explanation of the Repeating Patterns Index Question

Ms. Smith opted to use an index question as the preassessment. The index question includes two tasks. The students need to continue the pattern shown in Figure 3–9 and create the same type of pattern using letters. Clearly the first task is easier than the second task; this was intentional. Ms. Smith knows that students can often continue even complex patterns before they can describe patterns. She also knows that if she asks her first graders to describe the pattern with written words, many of the students will not be able to do this given that a majority of them are just learning how to write. *The preassessment*

Name _____

Leon drew this pattern, but he did not finish it.
Continue the pattern.
Draw the next four shapes.

☐ ◿ ◿ ☐ ◿ ◿ ☐ __ __ __ __

Use letters to make the same type of pattern that Leon made.

___ __ ___ __ ___ __ ___ __ ___ __ ___ __ ___ __

Figure 3-9 *Repeating Patterns Index Question*

See:
Number, p. 69
Data, p. 51
Geometry, p. 51
Measurement, p. 50

needs to focus on the math, not serve as a stumbling block for those students who are not yet writing. Making the same pattern with a different set of objects is more complex than continuing a pattern. Therefore the second task is a way to uncover who understands this type of repeating pattern at a higher level.

Student Responses

Shameika's Response

Take a look at Shameika's response in Figure 3–10. First, this is not an assessment of fine motor development. Ms. Smith is not focused on the fact that Shameika drew equilateral triangles rather than right triangles. What she sees is that Shameika can extend a pattern, but she is not yet able to create a new pattern using different items. However, Shameika's response is exciting because Ms. Smith has not started teaching repeating patterns yet, so the fact that some students already know how to extend ABB patterns is wonderful news. There were other students who gave responses similar to Shameika's. Ms.

Figure 3-10

Shameika's Response

Index question: Repeating patterns

Name Shameka

Leon drew this pattern, but he did not finish it.
Continue the pattern.
Draw the next four shapes.

Use letters to make the same type of pattern that Leon made.

Smith formed a group of these students and focused on teaching them how to describe patterns in ways that help them form similar patterns. By doing so, Ms. Smith targeted instruction to help Shameika and the rest of his group gain further knowledge of repeating patterns.

Sylvia's Response

Sylvia's response in Figure 3–11 indicates that she knows a lot about repeating patterns. Sylvia can extend a pattern and create the same pattern using letters. Sylvia already knows the on-grade-level concepts associated with repeating patterns. Ms. Smith understands that Sylvia and the other students who gave similar responses will need a challenge. They will need to work with higher-level patterns and more complex tasks. Ms. Smith does not want to waste the time of these students by making them sit through a whole group lesson on a topic that they already completely comprehend.

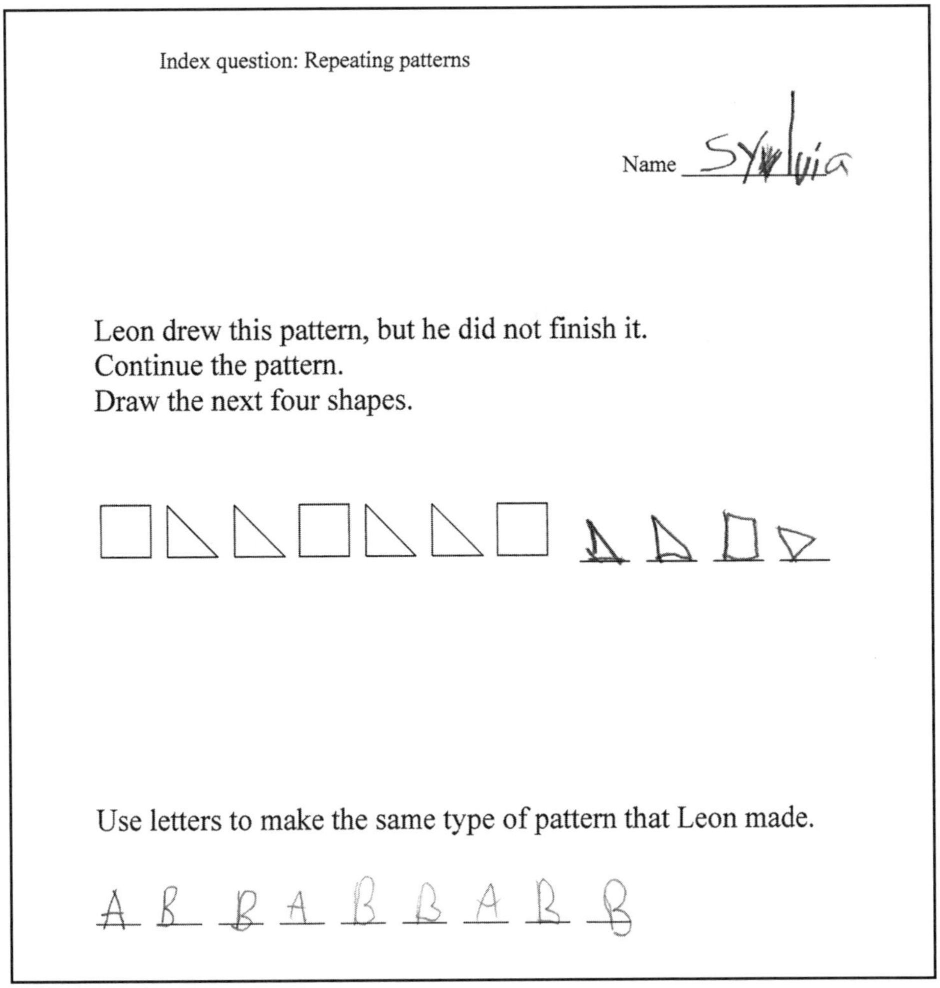

Figure 3–11
Sylvia's Response

Ever's Response

Ever's response in Figure 3–12 makes Ms. Smith smile. While the response indicates that Ever does not know either how to continue ABB patterns nor how to use letters to make the same type of pattern, he did complete the task! And he wrote his name! This is progress for Ever, who started the school year not knowing how to do either. While these skills are not directly related to repeating patterns, they are important foundational steps in learning how to communicate math thinking. Ever and the other students who gave incorrect responses will work in a small group with Ms. Smith. She will scaffold the math with these students by going back to simpler patterns and attributes.

Using a simple index question allowed Ms. Smith to form groups based on student needs. The preassessment took the students three minutes to complete. It took Ms. Smith about two minutes to do a quick sort of the indexes and form her purposeful groups for the day's instruction. Now Ms. Smith can

Figure 3–12

Ever's Response

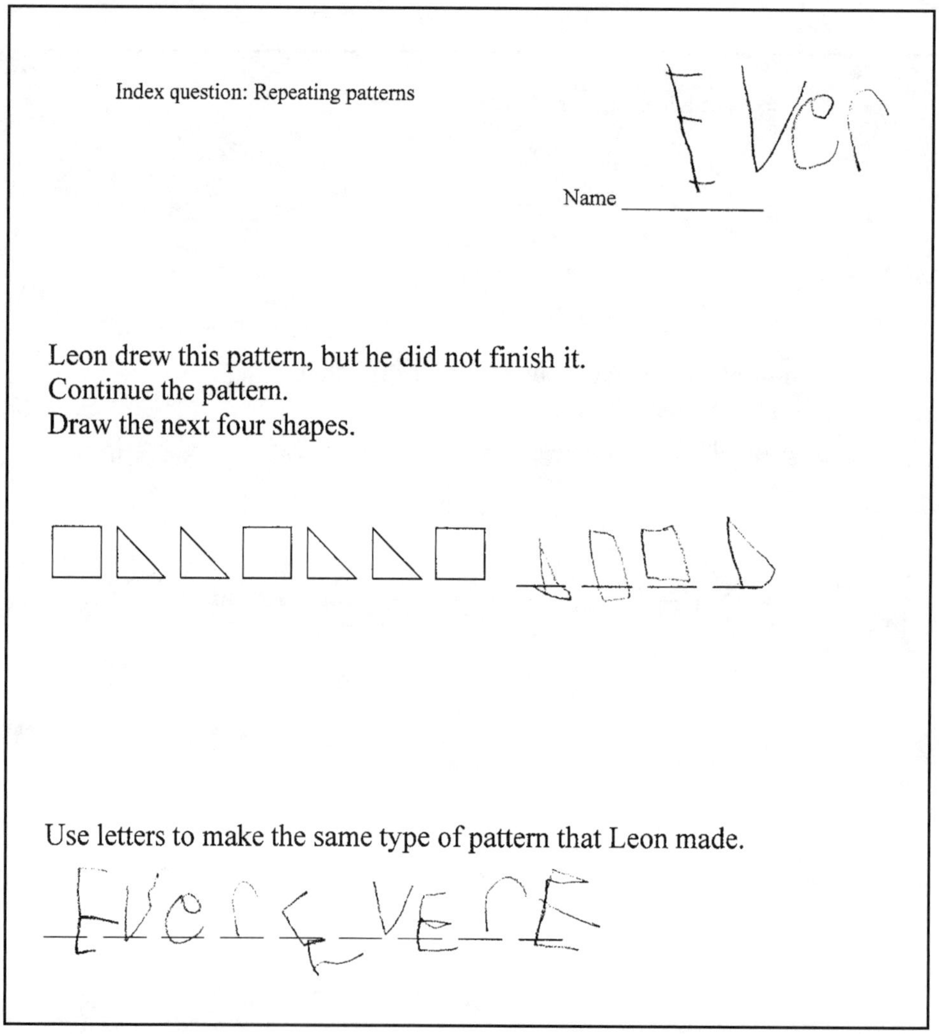

Index question: Repeating patterns

Name _____

Leon drew this pattern, but he did not finish it.
Continue the pattern.
Draw the next four shapes.

Use letters to make the same type of pattern that Leon made.

focus her attention on adjusting the levels of cognitive demand and targeting instruction aimed at meeting the students' academic needs.

Small Group, Targeted Instruction

See:
Number, p. 74
Data, p. 54
Geometry, p. 56
Measurement, p. 53

There were three distinct levels revealed in the responses given on the pre-assessment Ms. Smith gave to her first graders. Ms. Smith plans to meet with each group to target instruction.

The First Group Works with Repeating Patterns

The middle group (including Shameika) appears to have a solid understanding of how to extend ABB patterns, but does not yet understand how to make the same pattern with different objects.

This group needs to focus on describing patterns in ways that allow the students to show various forms of the same type of pattern. They do not need to spend time extending ABB patterns because they already know how to do so. Ms. Smith compacts the information on extending ABB patterns and concentrates instruction on describing patterns. The students use manipulatives to make repeating patterns and "read" the patterns as they point to and describe the objects in the pattern. Ms. Smith asks questions such as, "How would you describe this pattern?" "How do you know what comes next?" and "What is the repeating core of the pattern?" The students engage in mathematics discourse as they ask each other similar questions. Partners within the group are then given the task to take turns creating two repeating patterns and asking their partners if the patterns are same or different even though each pattern is made with different objects. Ms. Smith facilitates the dyads as they get started on the task. Students are asked to continue this task and record the patterns in their math journals.

The Second Group Works with Repeating Patterns

The responses given by this group of students (including Sylvia) indicate that they already know how to extend ABB patterns and how to make the same pattern using different objects. In essence, this group already knows how to do the on-grade-level material that the curriculum pacing guide states the class should do for the next two days—which would be a waste of time for these students. They need a challenge. Ms. Smith uses forward-mapping to determine that she needs to present repeating patterns beyond ABB types. Ms. Smith divides the small group into pairs and gives each pair a different set of manipulatives (shaded pattern blocks, colored cubes, counting bears). After writing

AABCAAABCAAABCA on the dry-erase board, Ms. Smith asks the students to show the same pattern using their manipulatives. Pairs are encouraged to describe the pattern as they make decisions about how to show it. All of the students complete the task quickly and accurately; therefore, Ms. Smith decides to incorporate a higher level of cognitive demand to add more challenge for the students. She presents the repeating pattern shown in Figure 3–13 to the students and asks them to show it using their manipulatives.

The students talk with their partners, but they are perplexed. They are not sure what comes next in the pattern, let alone how to show the same pattern with different objects. Ms. Smith encourages their thinking by asking them to describe what they see and explain why it is confusing. After some discussion, Ms. Smith asks, "What if all the shapes were the same color?" The students quickly respond that it would make the pattern a simple circle/triangle (AB) pattern. Ms. Smith continues, "How can you use that information to continue this pattern?" One of the students asks, "So does the color matter or not?" Ms. Smith helps the students use prior knowledge by asking, "If the color does not matter, what could come next in the pattern?" The students agree that a circle of any color could come next. Ms. Smith explains that in this pattern, color serves as a distracter and the only things that matter are shape and size. Ms. Smith encourages the students to try making a pattern that includes color as a distracter. One of the students working with the colored cubes announces, "We can't because we have all the same shape and size." The students working with the sorting bears agree, "We can't either because all we have are bears that are different colors." Ms. Smith replies, "What other attributes could you use to make the pattern?"

The students using the bears decide to position the bears differently. They create a pattern using random colors, but deliberate, repeated positions. They describe the pattern as "standing up, sleeping, standing up, sleeping, standing up, sleeping . . ." The students using the colored cubes use random colors, but group the cubes into deliberate, repeated amounts. They describe the pattern as "tower of two, tower of three, tower of two, tower of three, tower of

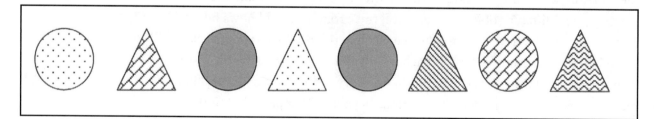

Figure 3–13 *A Higher LCD Repeating Pattern*

two, tower of three . . ." The students using the shaded pattern blocks create a pattern using random shades and random shapes, but deliberate, repeating positions. The simple repeating pattern that they created in Figure 3–14 was actually quite complex.

The students described the pattern in this way, "It is like the shapes are standing, flat, tippie-toe, flat, tippie-toe, flat, tippie-toe . . ." Ms. Smith used targeted instruction to help these students extend their level of knowledge related to patterning. Because the idea of using shading or color as a distracter is complex, Ms. Smith scaffolded the instruction by using more simplistic patterns (AB) to introduce the complex concept. Afterward, she gave the students the task of using color as a distracter while creating and describing patterns other than AB and she asked them to record these patterns in their math journals.

The Third Group Works with Repeating Patterns

These students (including Ever) gave responses that indicate that they are not yet able to extend ABB patterns nor do they know how to make the same pattern using different objects. Ms. Smith uses back-mapping and scaffolding with this group. She knows that working with easier patterns is the place to start, so she shows the students a simple AB pattern and asks them to describe how to extend it. By so doing, Ms. Smith encourages the students to focus attention on the preceding concept (simpler pattern type). Ms. Smith shows the students a pattern (AB) that includes two colors using counting chips. The students describe what they see and how to read the pattern. Ms. Smith asks, "How could we clap and tap this pattern?" The students respond and demonstrate an accurate AB pattern by clapping and tapping. Ms. Smith bumps up the level by changing the pattern (to an AAB) and asking the students to clap and tap the new pattern. All of the students clap and tap an AB pattern, not the AAB pattern. Ms. Smith addresses the error by asking them to compare the new pattern to the previous pattern. The students investigate the "red, red, yellow, red, red, yellow, red, red, yellow . . . " compared to the "red, yellow, red,

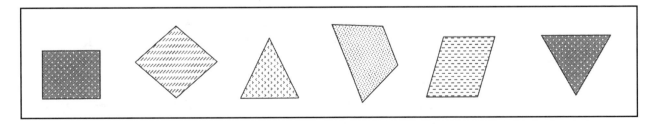

Figure 3–14 *A Student-Created Simple Repeating Pattern*

yellow, red, yellow . . ." Ms. Smith continues to scaffold the instruction by asking the students to connect their clapping and tapping to the corresponding colors. After successfully completing the tasks, the students are asked to try clapping and tapping shape patterns. Ms. Smith helps the students connect the new situation (shape pattern) to the previous situation (color pattern) and they learn to clap and tap AAB patterns. Ms. Smith then teaches them how to translate a clap and tap pattern into a pattern shown with manipulatives and depicted with letters. The students record the patterns in their math journals.

In this differentiated mathematics lesson, Ms. Smith targets the instruction for each group. This is not *the* way to differentiate instruction, but it is *a* way to differentiate instruction. There is not a single correct path, but rather many options when differentiating instruction. The path Ms. Smith chose was very successful because the instruction was targeted to meet the academic needs of the students. Ms. Smith also incorporated a tiered assignment for the students to work on after her direct instruction. She has the students record the specific tasks in their math journals for several reasons. She knows they need to represent patterns, but she also wants a record of what they work on so she can intermittently monitor what they are doing. During each of the small group targeted instruction sessions, she left the group she was working with for just a few minutes to check on what the other students were doing. Because the students recorded the patterns in their math journals, she could give a quick check to make sure they were not developing misconceptions due to inaccuracies. Ms. Smith used appropriate strategies to target the instruction for the students. All of the students progressed—everyone learned something related to repeating patterns.

Index Question for Equality and Balance

Equality and balance = balance and equality

See:
Number, p. 77
Data, p. 57
Geometry, p. 59
Measurement, p. 56

Mr. Hamman teaches second grade. He has twenty-five students in his class. Many of his students are ELL (English Language Learners). All of his students have diverse needs and various levels of mathematics understanding. Mr. Hamman is teaching a three-day unit on algebra. The algebra standard is "using and analyzing a variety of representations," specifically how equality and balance serve as cornerstones of algebraic thinking. After one day of instruction, Mr. Hamman decided to give his students an index question to help him assess the needs of his students.

The index question Mr. Hamman used (see Figure 3–15) included two balanced scales and one scale that needed to be balanced. Students were asked to use the information on the balanced scales to construct their own balanced scale. Additionally students needed to explain something about the process they used to construct a balanced scale. This index question requires students to use what they know about equality and balance to think algebraically. The first scale gives us important information. By using the notions of equality and balance to mentally remove one plaid box from each side of the scale, we can isolate the black box (later to be known as *x*) and reveal that the black box weighs the same as two spotted boxes. The second scale gives us less complex information. The plaid box is already isolated—it weighs the same as one black box and one spotted box. This index question will give Mr. Hamman information about who understands balance and equality as connected to algebraic thinking. The explanations students give will provide further information about the kind of thinking students are using to solve the math task.

Student Responses

Henry's Response

Considering that this assessment was given after some instruction on the topic, Henry's response in Figure 3–16 is interesting. He shows a solid level of understanding in his creation of a balance scale. Evidently he was able to use his understanding of equality to reveal that two spotted boxes weigh the same as one black box. However, he was unable to explain his thinking. He wrote *I gast*, which means he guessed. It is hard to tell Henry's exact level of understanding, but the fact that he was able to create a balanced scale from the complex information indicated to Mr. Hamman that Henry needed some challenge and further experiences with explaining the algebraic thinking associated with equality and balance. There were several other students in the class that were able to create a balanced scale, but were unable to accurately describe the math thinking. So Mr. Hamman grouped these ten students together and adjusted the level of cognitive demand associated with the task. Mr. Hamman targeted the instruction to the specific needs of the group.

See:
Number, p. 78
Data, p. 58
Geometry, p. 60
Measurement, p. 57

Andrea's Response

Andrea's response (see Figure 3–17) is extraordinary. Andrea is able to combine the information in both balanced scales to reveal the weight of the plaid box. Andrea's explanation is "One plaid box equals one black box plus one spotted box. One black box equals two spotted boxes. So, one black box plus

Index Question

Name _____

This scale is balanced.

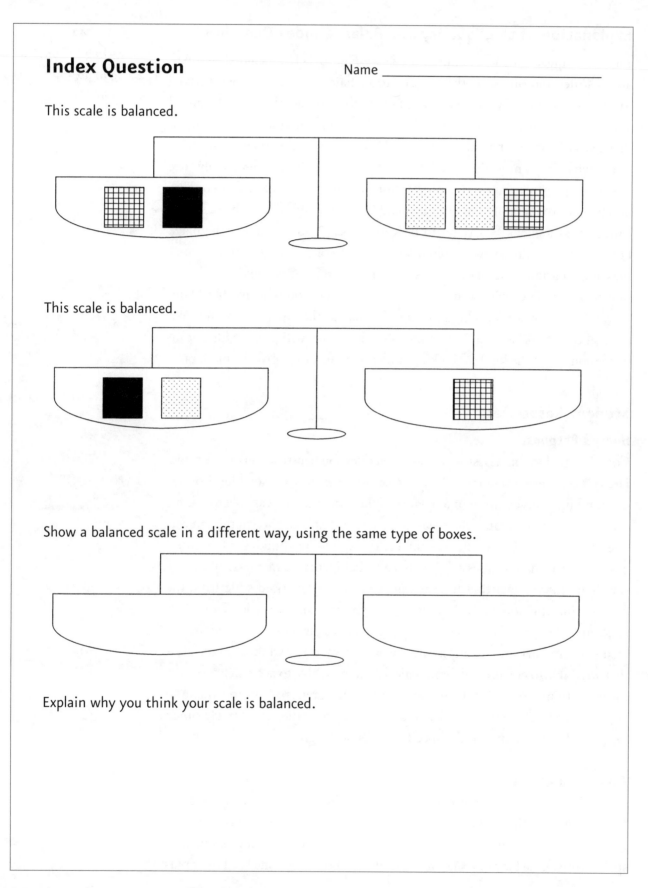

This scale is balanced.

Show a balanced scale in a different way, using the same type of boxes.

Explain why you think your scale is balanced.

Figure 3–15 *Equality and Balance Index Question*

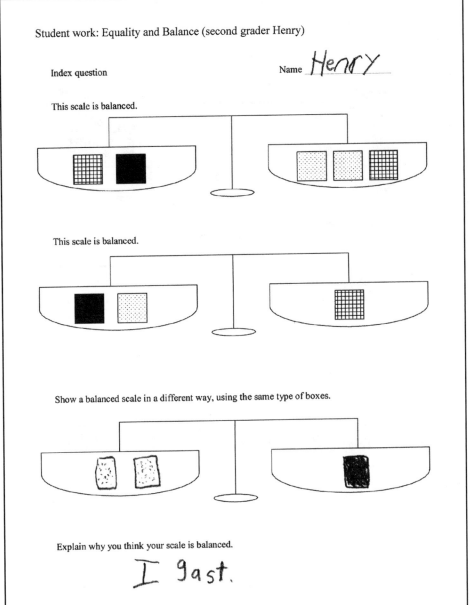

Figure 3–16 *Henry's Response*

one plaid box equals three spotted boxes plus one black box." Needless to say, Andrea is working at a very high level. She understands how to find the value of unknowns through the use of balance and equality. Mr. Hamman grouped Andrea with other students who responded with similar levels of understanding. Obviously Mr. Hamman needed to give this group a challenging task and targeted instruction aimed at a higher level.

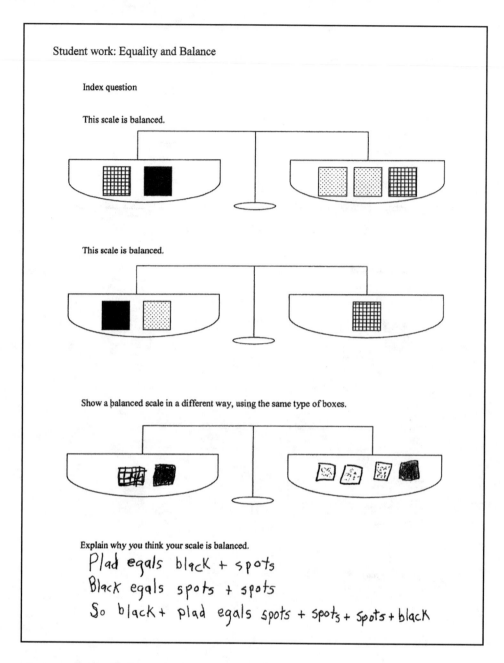

Student work: Equality and Balance

Index question

This scale is balanced.

This scale is balanced.

Show a balanced scale in a different way, using the same type of boxes.

Explain why you think your scale is balanced.

Plad eqals black + spots
Black eqals spots + spots
So black + plad eqals spots + spots + spots + black

Figure 3–17 *Andrea's Response*

Carly's Response

Carly's response shown in Figure 3–18 includes some accurate math ideas, but the ideas are not directly related to the task. "I know that [my scale is balanced] because 3 + 3 = 6 and 6 is equal and there's 6 altogether" is in fact true. However, Carly does not use the other balanced scales to reveal information about the comparative weight of the different boxes. She has some understanding of balance because both sides of the scale are the same. But her level of under-

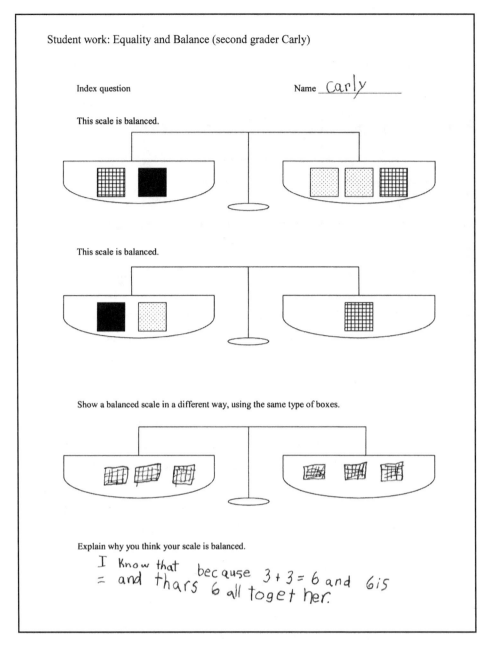

Student work: Equality and Balance (second grader Carly)

Index question   Name Carly

This scale is balanced.

This scale is balanced.

Show a balanced scale in a different way, using the same type of boxes.

Explain why you think your scale is balanced.

I Know that because 3 + 3 = 6 and 6 is = and thars 6 all together.

Figure 3–18 *Carly's Response*

standing is not high. Carly and the other students who gave similar responses needed some additional work with simple notions of balance and equality. Mr. Hamman needed to provide some scaffolding to help this group gain a more solid foundation.

Mr. Hamman's use of the index question allowed him to reveal the levels of understanding of his students. The index question was not a preassessment, rather it was an assessment during a three-day unit on one component of algebraic thinking. It only took the students a few minutes to respond to the index question and just a few more for Mr. Hamman to quickly sort the information, which helped him form his groups for targeted instruction.

Small Group, Targeted Instruction

See:
Number, p. 81
Data, p. 62
Geometry, p. 62
Measurement, p. 61

There were three separate levels revealed in the responses given on the index question Mr. Hamman gave his second graders.

The First Group Works with Equality and Balance

The middle group (including Henry) appears to have a solid understanding of how to create a balanced scale, but does not yet understand how to explain the thinking behind it. Because this group understands how to show balance, the focus of the instruction should be on how to describe balance and equality. Mr. Hamman presents a balanced scale with weighted cubes to the small group and asks, "How can we describe the balanced situation?" The group does not respond. Mr. Hamman takes out a few weighted cubes to make the balanced situation less complex. The students are able to describe this balanced situation with words such as *equal, same,* and *even.* Continuing to scaffold the instruction, Mr. Hamman encourages the students to use the idea of addition as they describe each side of the balanced scale. After they successfully describe the simple balanced situation, Mr. Hamman adds more weighted cubes and shows an unbalanced scale. The problem posed to the students is to think of how to describe what needs to be done to make the scale balanced. The students engage in math discourse prompted and guided by Mr. Hamman. The students are given a worksheet that has three empty scales and asked to work in dyads to create balanced scales with explanations.

The Second Group Works with Equality and Balance

This group of students (including Andrea) gave responses that indicate that they need a challenge. They already understand how to create and describe balanced situations. They understand equality at a level beyond the rest of the class. Mr. Hamman uses forward-mapping and compacting with this group. He compacts the instruction by having the students skip showing equality on scales because they already know how to do so. Instead he presents an equation containing unknown variables ($\Box + \Box + \Box = \Box + 6$) and asks the students to work in dyads to solve it. After the students talk it through, they share their answers with the group. Mr. Hamman then teaches the students how this problem can also be expressed as $3x = x + 6$. He helps the students connect to their prior knowledge of the relationship between multiplication and addition. He also opens the door for understanding by bridging from the known (solving problems with "☐") to the unknown (solving problems with "x"). The group works with several other similar problems.

The Third Group Works with Equality and Balance

These students (including Carly) gave responses that indicate that they are not yet able to show or explain a balanced scale. Mr. Hamman targets the instruction for this group of students by using some of the same techniques he used with the previous group, but he incorporates specific back-mapping and more scaffolding. He shows a simple balanced scale and encourages verbal descriptions. But this group is not able to come up with any math words or ideas. So Mr. Hamman gives the students the math words (*equal*, *balanced*, *same*, *even*, *add*, *plus*, *more*, *less*) and asks them to show each of the concepts with the scales and weighted cubes. After showing and discussing each of the concepts, the students are asked to draw and write about each concept. By showing each of the math ideas first, the students gain a better understanding of the concepts and are able to apply the concepts of balance as evident in their drawing and writing.

In this differentiated mathematics lesson, Mr. Hamman successfully challenges and supports each of the groups through targeted instruction. Mr. Hamman had many options for how to differentiate the instruction. He chose to tier the lesson, use back-mapping, use forward-mapping, scaffold, compact, and activate prior knowledge. All of the students gained further knowledge of equality and balance.

CHAPTER 4

Grouping Students in Algebra

Using Algebra Vocabulary to Label Small Groups

See:
Number, p. 98
Data, p. 67
Geometry, p. 69
Measurement, p. 67

Using content vocabulary to identify small groups offers a twofold benefit. The group names help with classroom management and offer opportunities for exposure to content vocabulary. Because the groups are always changing, we want the names of the groups to change as well. Using specific content vocabulary as group names encourages the students to see and hear more math. These are lists of some of the algebra vocabulary that could be used as names of small groups.

Patterns, relations, and functions words:

Ascending	Linear
Attribute	Pattern
Concentric	Predict
Descending	Recognize
Describe	Relationship
Extend	Repeat
Function	Repeating pattern
Generalize	Rule
Growing pattern	Translate

Representing and analyzing math situations words:

Associative	Equality
Balance	Operations
Commutative	Properties
Compose	Situation
Decompose	Structure
Equal	Symbol

Using models to represent quantitative relationships words:

Equation
Expression
Model
Quantitative
Relationship
Represent

Analyzing change words:

Change
Generalize
Patterns
Prediction
Qualitative
Quantitative

CHAPTER 5

Levels of Cognitive Demand in Algebra

Adjusting LCD in Problem Solving

A group of first graders works with the concept of equality using pan balances and cubes. Because the group has some difficulty with concept, the teacher uses a lower level of cognitive demand (LCD) by modeling ways to solve the learning situation. The teacher places six red cubes and four green cubes in one pan (all cubes are the same weight and size). Seven blue cubes are placed in the other pan. The teacher asks, "Are the pans equal?" The students respond, "No." The teacher asks, "What should we do to make the pans equal?" The students are not sure what to do. The group is quiet. The teacher asks, "Do you want to put some more cubes in or take some cubes out?" while demonstrating how to move the cubes and how to check the scale. The students take the teacher's lead and begin trying various ways to make the pans equal. They add three yellow cubes to the second pan as one solution. They take out three red cubes from the first pan as another solution. They remove all but two green cubes in the first pan and two blue cubes in the second pan as a third solution. By removing and adding cubes the students come up with many different accurate responses. The teacher adjusted the LCD by modeling a strategy and by asking a question that guided the students into thinking about the problem-solving options. In this scenario, the teacher did not require an extremely low LCD because the students were not told step-by-step what to do. There were also several possible answers to this problem solving situation. Yet the LCD was not extremely high because the students were given some modeling and suggestions. The teacher used the LCD that best fit the needs of the students.

Given the same topic (equality) but a different group of students, the teacher may differentiate the problem solving situation in another way.

This group of students understands the basics of equality and is ready to work on problem solving in a way that uses a much higher LCD. The teacher places twelve cubes in front of the students and poses this question, "How many times do we need to use the scale to prove that all the cubes are equal in weight?" The manipulatives (pan balances and cubes) are the same ones used by the other group, but the problem is different. Notice the students' thinking and the role of the teacher in the following dialogue:

Alvin: You have to put one cube in one pan and compare it to the rest one at a time [Alvin places one cube in each pan].

Maria: So how many times will we use the pan balance?

Angela: That would be twelve times.

Alvin: Let's see [Alvin weighs each cube against the first cube].

Jeremy: I'll count the times [Jeremy counts each use of the balance].

Maria and Angela: It's eleven.

Teacher: Why is it eleven?

Angela: It is like the first one doesn't count.

Jeremy: Because you can't weigh the first one against itself.

Alvin: So it is 12 − 1.

Teacher: Could you use the balance less than eleven times?

Alvin: No, because you have to weigh all of them to really prove that they are all equal.

Maria: What if we put two in each pan?

Jeremy: Let's try it [Jeremy places two cubes in each pan].

Alvin: I'll count this time.

Angela: I think it will be six times.

Teacher: Why do you think six?

Angela: Because six is half of twelve.

Alvin: But the answer is five.

Maria: It is one less than we thought it would be.

Jeremy: Just like last time. You can't weigh the first pair against itself.

Teacher: Is there a way to use the balance less than five times?

The group continues working with the pan balance and cubes, finding multiple ways to weigh the cubes in groups and eventually determining that they could use the balance just one time to prove that all the cubes are equal weight by putting six cubes in each pan. The problem solving situation was open-ended and there were several correct answers. Notice how the teacher did not model how to solve the problem, yet constantly raised the LCD by asking questions that helped the group continue to probe deeper into the math situation.

The levels of cognitive demand required in math problem solving situations should vary based upon the needs of the students. Adjusting the LCD based on the immediate academic needs of the students is an excellent way to differentiate mathematics instruction.

Adjusting LCD in Reasoning and Proof

See:
Number, p. 110
Data, p. 73
Geometry, p. 76
Measurement, p. 72

First graders Erin and Marcus are creating ABA patterns with shapes. The repeating pattern core is triangle, circle, triangle. The core of the pattern is repeated three times and Marcus correctly places a triangle next. The teacher asks, "Marcus, how do you know that the triangle comes next?" Marcus says, "My brain told me." This response (albeit adorable) does not shed light into Marcus' thinking. He has not yet explained how he knows. The teacher follows up with another question, asking, "What did your brain tell you?" Marcus says, "It told me to put a triangle next." The persistent and talented teacher continues, "Why didn't your brain tell you to put a circle next?" Marcus responds, "Well, it can't be circle, because the pattern is triangle, circle, triangle. If you put a circle, it will mess it up." The teacher asks, "Why?" Marcus replies, "It has to go again and again the same way." Marcus has now moved to using reasoning to explain what he is thinking. The teacher adjusted the LCD by offering questions that built upon each other. At first, Marcus could not explain his reasoning. However, the teacher guided him with a series of questions that eventually resulted in Marcus being able to explain what came next in the repeating pattern.

First graders Jack and William are creating ABA patterns with shapes. The core of the repeating pattern is rectangle, triangle, rectangle. The pattern core is repeated twice. The teacher asks, "May I add some shapes to your pattern?" The children nod enthusiastically. The teacher places a square, followed by a triangle, followed by a nonsquare rectangle. The boys frown. William says, "The square is not right." The teacher asks, "Why not?" Jack explains, "We are only using rectangles and circles." The teacher replies, "Are squares part of the rectangle family?" Both boys reply, "Yes," but they look a bit perplexed. The teacher says, "Think about how you could read this pattern (with the square in it) so that it is a true repeating pattern." After some thought, Jack announces, "Four sides, three sides, four sides, four sides, three sides, four sides." William joins in pointing at the square, "*Four sides*, three sides, four sides, because squares and [nonsquare] rectangles have four sides." The teacher asks, "Why did you decide to count the sides?" Jack says, "I just kept asking myself how the rectangle and square are the same because if they have the same place in the repeating pattern, they have to be the same in some way." William adds, "You could just say what family the shapes are in, like rectangle, triangle, rectangle, rectangle, triangle, rectangle, *rectangle*, triangle, rectangle—because the square is in the rectangle family." In this situation, the teacher increases the LCD required by adding complexity to the situation while asking probing questions. The students use geometric and algebraic reasoning and proof to explain and justify the repeating pattern.

Adjusting LCD in Communication

Sometimes students communicate algebraic concepts that they do not yet understand. For example, many students do not understand the concept of equality, yet they use the word *equal* and the equal sign on a regular basis. Similarly, students may not understand how to communicate computational concepts. I once asked a first grader how his addition was coming along. He replied, "I'm not doing addition, I am plussing."

Communication provides another way to adjust the LCD as we differentiate instruction in mathematics. Of course, we want all children to communicate in mathematics, but we can use the various types and levels of sophistication to adjust the levels of cognitive demand.

The teacher works with small groups of kindergartners on repeating patterns. The students use soft foam shapes to make the repeating patterns. The teacher adjusts the LCD in communication. With one group, she asks the students to describe and explain with words as they show models of repeating

See:
Number, p. 112
Data, p. 74
Geometry, p. 76
Measurement, p. 74

patterns. The students' descriptions involve "reading" the patterns. Joseph shares, "Blue triangle, red circle, blue triangle, red circle, blue triangle, red circle." Chaya shares, "Green square, yellow triangle, green square, yellow triangle, green square, yellow triangle." The teacher invites the students to "read" one another's patterns. The LCD in communication is relatively low in this situation.

With another group of kindergartners, the teacher uses a slightly higher LCD. These students use the same manipulatives (soft foam shapes) to model repeating patterns. The teacher asks, "What comes next?" and "How do you know?" Sondra replies, "The next one in my pattern is a rectangle. I know because it goes rectangle, not rectangle, rectangle, not rectangle, rectangle, not rectangle." The teacher encourages the students to take turns asking one another questions about the patterns. Sondra asks Grady, "What comes next in your pattern?" and "How do you know?" After Grady answers, he asks another student the same questions. The process continues until all of the students have asked and answered the questions. Even though the level of thinking is not extremely high, the LCD in communication is at a higher level because the students are asking each other questions.

Another group of kindergartners requires a much higher LCD in communication. With this group, the teacher encourages the students to analyze the shape patterns by communicating the similarities and differences between the repeating patterns. Benjamin shares, "My pattern is like Theo's pattern because we both used three things to make our patterns." Thresher shares, "My pattern is different from Andy's pattern because I used circles in my pattern and he didn't use any circles." Xuan announces, "My pattern is not like anyone else's." However, CJ disagrees, responding, "Well, it is sort of like mine." Xuan asks, "How?" CJ replies, "Mine is three corners, four corners, three corners, four corners, three corners, four corners, and so is yours!" After studying the two patterns, Xuan exclaims, "You are right. But it is weird because I used different shapes." The LCD in communication is high in this situation because the students are analyzing and challenging each other's thinking.

Adjusting LCD in Connections

See:
Number, p. 117
Data, p. 75
Geometry, p. 81
Measurement, p. 76

Algebra is connected to many topics outside of mathematics, such as science, art, literacy, social studies, physical education, music, and technology. Helping students make these connections enhances their understanding and application of algebraic thinking. Algebra is also connected to all of the other math content standards in multiple ways. The ties between algebra and number sense are especially profound. Using math models to represent and understand quantitative relationships, for example, is an algebra concept that is completely embedded in number sense. Other connections within the standards may appear less direct until one takes a closer look. Geometry, for example, may seem like a very different field of study when compared to algebra. Yet many patterns involve shapes and positions. Other patterns may involve measurement. Likewise, sets of data may include patterns and manifestations of quantitative change. Helping students understand these and other connections can serve as a way to adjust the LCD when differentiating instruction in mathematics.

In a nutshell, if the connections are simple, the LCD is low. If the connections are complex, the LCD is high. Teachers need to adjust the levels based on the specific needs of the students.

A class of second graders is working on analyzing change in various contexts. The teacher has preassessed the students and learned that some need a higher LCD and others need a lower level. Yet the teacher plans for all of the students to grow in their understanding of algebraic change and making connections.

The class is given this prompt: *How and why did we use math in yesterday's science experiment?* All students are to address this question by talking to their science partners, writing independently in their math journals, sharing what they have written with their partners, and adding clarifying illustrations to the math writing. While the students are writing, Ms. Jeffries meets with six students to provide extra support. Ms. Jeffries displays the plant growth data and asks, "Let's look at this particular plant [Ms. Jeffries chooses a plant that has shown steady growth]. How would you describe the change?" The students discuss the changes that have occurred. They decide that the height of the plant has increased over time, but they are not sure about the exact increase. Ms. Jeffries tells the children to "use linking cubes to show the height of the same plant each day." The students begin working on the task. Ms. Jeffries helps the students connect algebra, measurement, and science by asking questions and guiding the students through the task. After the task is completed, the students go back to writing in their math journals.

Before Ms. Jeffries meets with the next group, she walks around the classroom checking on the other students' math writing and sharing progress. She offers suggestions, answers questions, and redirects a couple of students. Ms. Jeffries now calls the next group to work with her at the back table. This group has a strong understanding of algebraic change. She offers the group a task that requires a higher LCD by posing more sophisticated connections. Ms. Jeffries asks, "What do you notice about the data?" The students discuss the changes that have occurred. Ms. Jeffries gives the following task: "Construct two bar graphs that show the change over time. The first bar graph will be of the fastest growing plant. The second bar graph will be of the slowest growing plant." The students begin working on the task. Ms. Jeffries helps the students connect algebra, measurement, data organization, data display, data analysis, data interpretation, and science by asking questions and guiding the students through the task. Ms. Jeffries teaches the students how to find the average growth using the fastest and slowest growth data. After the task is completed, the students go back to writing in their math journals.

Ms. Jeffries meets with the third group of second graders who showed that they have a mid-range understanding on the preassessment. Ms. Jeffries uses a combination of the two previous tasks for this group. The students use the plant growth data to make towers out of the linking cubes. Ms. Jeffries teaches the students how to use the towers to make a three-dimensional bar graph as they learn more about quantitative change and making connections. The LCD required is not too high and not too low—just right for this group of students.

Adjusting LCD in Representations

See:
Number, p. 114
Data, p. 78
Geometry, p. 83
Measurement, p. 79

Second graders are learning about commutativity. Ms. Salem adjusts the LCD in representation as she works with two different groups of students. Some of the students have some knowledge of the commutative property. When Ms. Salem meets with this group of students, she writes four problems on the dry-erase board:

$$19 + 16 =$$
$$32 - 14 =$$
$$2 \times 3 =$$
$$10 \div 2 =$$

She asks, "Does the order matter?" One student says, "Yes." Another student says, "It depends." The students engage in a discussion about addition and multiplication having the order property and subtraction and division not having the order property. Ms. Salem invites the students to represent their ideas about the order of operations. The students decide to write on number lines to represent the commutative property for the addition and the multiplication equation. After some discussion, they decide that they can also use number lines to represent why subtraction does not have the commutative property. One of the students explains, "The difference between 32 and 14 is 18. It is really neat that the difference between 14 and 32 is also 18. But it is not the same because it is the 18 that comes before zero." Another student adds, "Yes. That's negative 18." The students were a bit perplexed about how to represent 2 divided by 10, but they felt confident that it is not the same as 10 divided by 2. Ms. Salem asks, "If you have 2 cakes and 10 people what would you do?" The students decide to cut two pieces of paper into ten equal parts to represent the equation. One of the students suggests that they also show ten pieces of paper divided into two equal parts to represent the original equation.

Ms. Salem requires a higher LCD for this group of students. She invites them to think about how to represent their thinking. Even when they are not sure how to represent that division does not have the commutative property, she does not give them the answer. Instead, she asks a question that places the mathematics in context, which prompts the students to think of an appropriate way to represent the idea.

Some of the other students in Ms. Salem's class are not very comfortable with the commutative property. When Ms. Salem meets with this group, she adjusts the LCD to a lower level. She invites the students to think about 4 + 8. She asks if it is the same as 8 + 4. The students are not sure. Ms. Salem has the students build a tower representing 8 + 4 using 8 red cubes and 4 yellow cubes. Then she asks the students to build a tower representing 4 + 8 using 4 red cubes and 8 yellow cubes. The students stand the towers side by side. Ms. Salem asks, "What do you notice?" The students reply, "They are the same." Ms. Salem picks up one of the towers and says, "8 + 4 is the same as . . ." then she flips the tower and says "4 + 8." Ms. Salem asks, "What about 15 + 6?" The students build the towers to represent the expression. After several more addition expressions, Ms. Salem asks the students, "Is 12 − 4 the same as 4 − 12?" Many of the students answer "Yes." Ms. Salem represents the idea with cubes. She shows 12 cubes and takes away 4 cubes. Then she shows 4 cubes and says, "Take away 12?" The students are not sure what to do. Ms. Salem says, "If I owed you 12 dollars and all I had were 4 dollars, I could pay you the four I have; but how many would I still owe you?" The students think about this and announce, "You owe 8 dollars." Ms. Salem shows the representation on a number line and prompts a discussion about the difference between an owed 8 (negative 8) and a paid 8 (positive 8).

In this situation Ms. Salem adjusts the LCD with more concrete representations. The work is more teacher-directed and less open-ended.

See:
Number, p. 125
Data, p. 79
Geometry, p. 84
Measurement, p. 81

Using the Levels to Differentiate Math Tasks

Let's look at one algebra concept, *equality*, and see how it can be adjusted to the various levels of cognitive demand.

Level 1 task: Define *equality*

Level 2 task: Explain what *equal* means

Level 3 task: Show how to make the groups equal

Level 4 task: Classify the expressions

Level 5 task: Recommend a way to teach equality to someone else

Level 6 task: Create a new symbol for equal

Using these tasks to differentiate instruction does not mean that all students will proceed through the tasks in consecutive order. Some groups of students may work on the first couple of levels, others on the middle levels, and yet others may only work on the higher-level tasks. The important thing is that the children are given tasks that are appropriate for their particular level.

The levels of cognitive demand offer many windows of opportunity to differentiate instruction in mathematics. I am always in search of new ways to *bump up* the LCD or *tone down* the LCD to enable my students to better grasp the algebra that I am teaching. Ultimately my goal is for all students to receive just the right amount of support and challenge as they learn and grow.

CHAPTER 6

Learning Frameworks in Algebra

Incorporating Learning Styles

A prekindergarten class begins a unit on sorting. The teacher understands that sorting is an important algebraic concept because it serves as a prerequisite of recognizing and extending repeating patterns. The teacher has assessment data that indicate the students' learning styles.

See:
Number, p. 128
Data, p. 80
Geometry, p. 85
Measurement, p. 83

Visual Learners Sort

When the teacher meets with the students who are visual learners, she provides pictures of actual objects (hammer, nails, saw, bike, yo-yo, and doll) for the students to sort. The teacher uses eye contact with each child. During instruction, the teacher incorporates body language and facial expressions to help the children connect with the instruction. The teacher has the students sort the items on large paper so that they can draw rings around the groups to help the students see the sorting decisions.

Auditory Learners Sort

When the teacher meets with the auditory learners the focus is on sound, pitch, and voice. The students listen to recorded noises and decide which sounds will be grouped together. There are automobile sounds (horn blowing, screeching tires, and motor revving) and animal sounds (dog barking, cow mooing, and birds chirping). The teacher encourages the students to imitate the sounds and to talk about the noises that they hear. Guided by the teacher, the students create a song that describes how they sorted the sounds.

Kinesthetic Learners Sort

The kinesthetic learners meet with the teacher in the front of the room. Each child is seated around a large rug. In the center are two hoops, one is labeled *Things made of paper* and the other is labeled *Things made of metal*. The students are invited to search around the classroom to find items to go in each group. The teacher helps the students transport the items back to the large rug and place each item in the correct location. After the task is completed, the students stand around the hoops and take turns dramatizing how they carried each item to its place in the sorting hoops. The movements the students make highlight some of the attributes of the objects (e.g., heavy, wide, delicate).

Tactile Learners Sort

The tactile learners work with the teacher at the back table sorting leaves. They move leaves into groups and decide how they will label the groups. The teacher encourages the students to feel the leaves as they describe the attributes (bumpy and smooth). The students point to and pick up the leaves as they use hands-on experiences to analyze their sorting decisions.

In this prekindergarten example, the learning style preferences were addressed in ways that helped the children better understand how to sort. The instruction was differentiated to meet the learning needs of the students. Additionally, each group created a final product that reflected the learning style preference—the visual group created a chart, the auditory group created a song, the kinesthetic group created dramatic interpretations, and the tactile group created a three-dimensional sorting diagram.

Incorporating Multiple Intelligences

See:
Number, p. 132
Data, p. 82
Geometry, p. 86
Measurement, p. 86

In a second-grade class the children are learning about algebraic reasoning as they represent and analyze math situations. The class has had many experiences with equality and equations. The students have participated in a recent inventory of multiple intelligences and the teacher grouped students by likenesses in intelligences. During the lesson the teacher serves as a facilitator, moving from group to group answering and asking questions, advising, assisting, and directing.

All groups are given the same three balanced situations shown in Figure 6–1 and asked to solve the last one. However, each group is asked to solve the problem (*showing the knowing*) in a different way.

61

Learning Frameworks in Algebra

Showing the Knowing in the Word Smart Group

Three students show strengths in word smart (verbal-linguistic intelligence). Their task is to create and tell a story about the algebra problem. The students discuss the balance situations using algebraic reasoning. Here is the story the group creates, revises, and eventually tells to the class:

> Once upon a time there was a weird family. The dad is a smiley face. The mom is a heart. The kids are bears and frogs. The family likes to stand on the balance. Two bears and one dad (smiley face) equals one mom (heart) and one dad (smiley face). If the dads (smiley faces) jump off the balance, then we know two bears equal one heart. The end.

Showing the Knowing in the Number Smart Group

Four students are high in number smart (mathematical-logical intelligence). These students work on assigning numbers to the symbols (bear, smiley face, heart, and frog) and creating equations. The students use algebraic reasoning to create associated equalities.

Bear is 2. Heart is 4. Smiley face is 1. Frog is 3.

$2 + 2 + 1 = 4 + 1$

$2 + 3 = 3 + 1 + 1$

$3 + 4 = 2 + 1 + 4$

The answers: $2 + 2 = 4$ or $1 + 1 + 1 + 1 = 4$

Showing the Knowing in the Picture Smart Group

There are five students who indicated some strengths in picture smart (visual-spatial intelligence). These students were given the task of making pictures of the symbols and creating a diagram. The students drew pictures (bear, smiley face, heart, and frog) on separate sticky notes and manipulated these as they used algebraic reasoning to solve the problem. They developed the following chart:

$B \, B \, S = H \, S \rightarrow B \, B = H$

$B \, F = F \, S \, S \rightarrow B = S \, S$

$F \, H = B \, S \, H \rightarrow F = B \, S$

Answer: $B \, B = H$

The first three balances are equal.

How can you make the last balance equal?

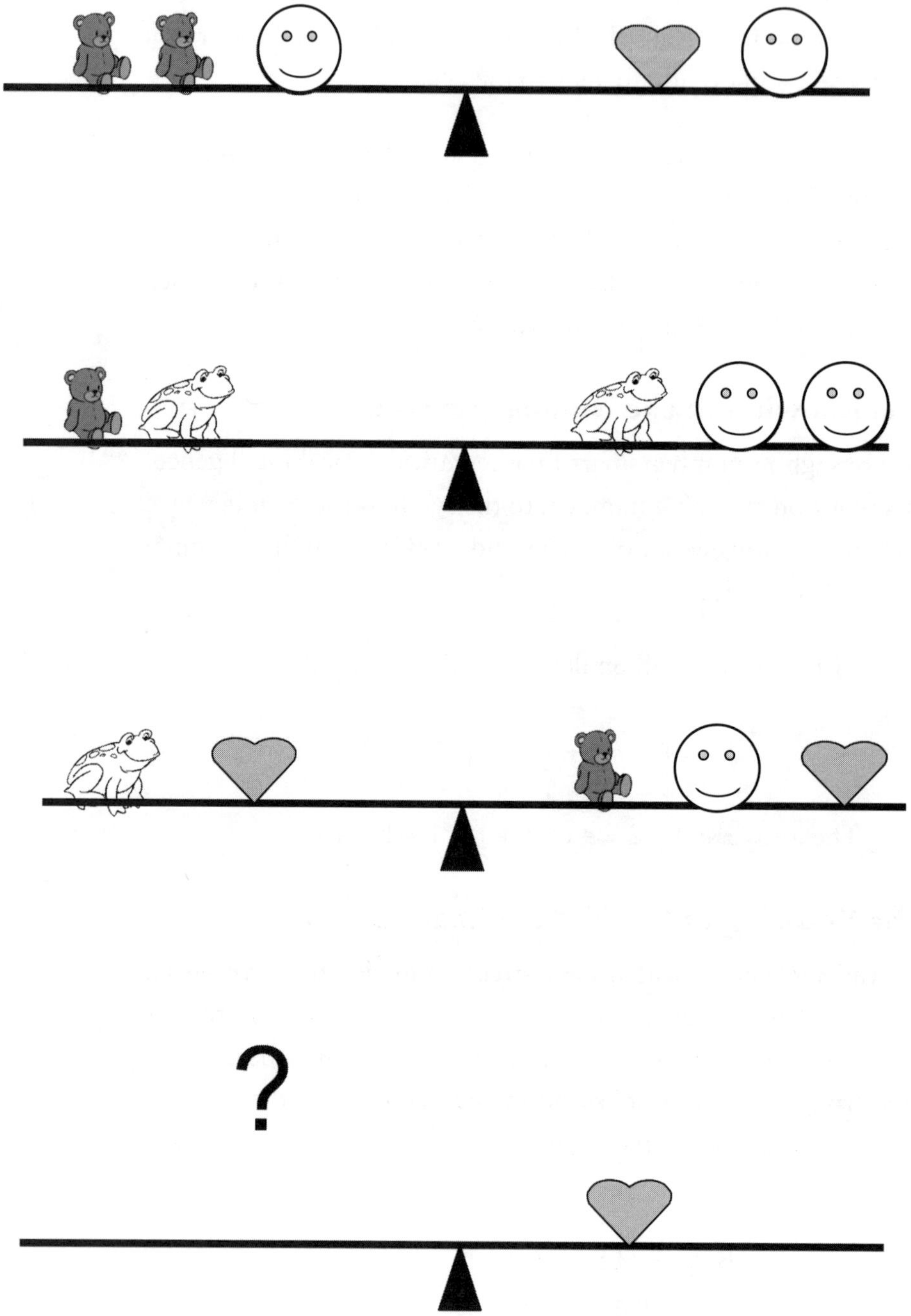

Figure 6–1 *Three Situations That Can be Solved to Satisfy Multiple Intelligences*

Showing the Knowing in the Body Smart/People Smart Group

There were three body smart (body-kinesthetic intelligence) students and two people smart (interpersonal intelligence) students who were grouped together and given the task of constructing a balance with people to represent and solve the math problem. These five students presented their dramatic interpretation of the problem by pretending to be the objects on the balance (bear, smiley face, heart, and frog). They used algebraic reasoning as they explained each balanced situation. They concluded that "one heart equals two bears or four smiley faces."

Showing the Knowing in the Music Smart Group

The final group consisted of mostly music smart (musical-rhythmic intelligence) students. These students created a rap song to describe the information and solution of the math problem. The sound effects were entertaining, to say the least. The words of the song were as follows:

> One side is a bear and a bear and a smile.
> The other side is a heart and smile.
> What does that mean?
> It means heart is the same as two bears.
> How did we figure it out?
> 'Cause we know about balance.
> 'Cause we know about equal.
> 'Cause we know about math!

In this second-grade example, the multiple intelligences were used as a means of *showing the knowing*. The students demonstrated understanding via their personal strengths. The instruction was differentiated to meet the learning needs of the students because the tasks and products were matched to specific intelligences.

Incorporating Environmental Needs

In a first-grade classroom the teacher strives to differentiate mathematics instruction based on environmental preferences. While studying patterns, a group of students learn that skip counting represents growing patterns. The students want to show this connection on the classroom number line. But the number

See:
Number, p. 138
Data, p. 85
Geometry, p. 88

line is placed on the wall close to the ceiling. The teacher adjusts the environment for the students by moving the number line to a place where the students can more easily interact with it.

Incorporating Affective Needs

See:
Number, p. 141
Data, p. 85
Geometry, p. 89
Measurement, p. 89

Jonathan and Jason are creating a growing pattern using cubes. Jonathan makes each new term in the pattern without consulting Jason or allowing him to put the cubes together. The teacher sees what is happening and invites Jason to use an *I message* to share his feelings with Jonathan. After some thought, Jason says, "I feel sad when you don't let me do any of the pattern because I am left out." Jonathan replies, "Oh. You do the next one and we will take turns." In this situation, the children were able to make the adjustment needed. Prior to the *I message*, Jason was not actively learning. After the *I message*, Jason's role became more dynamic. Jason and Jonathan worked together to construct and discuss the growing pattern. In this way, the *I message* prompted a change in the learning situation based on the affective needs of Jason, highlighting yet another way to differentiate mathematics instruction in the prekindergarten through second-grade classroom.

Incorporating Interests

In a kindergarten and first-grade combination class the teacher incorporates interests as she differentiates the mathematics instruction. The algebraic concept of study is using models to represent mathematics situations.

See:
Number, p. 147
Data, p. 86
Geometry, p. 89
Measurement, p. 89

The teacher groups the students in dyads with similar interests. While the dyads are homogeneous by interest, they are also heterogeneous by ability. Therefore the content is at the same level, but the instruction is differentiated via students' interests. The teacher poses this problem:

_____ , _____ , _____ , _____ , _____ (five friends)

met at _____ (setting).

Each person brought some _____ (collectable items).

There were twelve _____ (collectable items) in all.

How many _____ (collectable items) did each person bring?

Use manipulatives and/or pictures to show a model of your answer.

The students work together to insert personal interest details, creating a high-interest problem to solve. Two students, Erika and Marcia, create and solve the problem in this way (see Figure 6–2).

Erika, *Marcia*, *Jasmania*, *Britney*, and *Tasha* met at *the mall*.

Each person brought some *bracelets*.

There were twelve *bracelets* in all.

How many *bracelets* did each person bring?

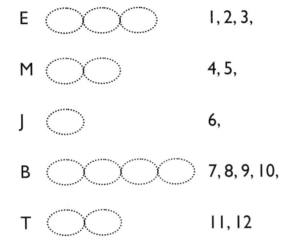

E	⬭⬭⬭	1, 2, 3,
M	⬭⬭	4, 5,
J	⬭	6,
B	⬭⬭⬭⬭	7, 8, 9, 10,
T	⬭⬭	11, 12

Figure 6–2 *Erika and Marcia's Response*

Alex and Danielle create and solve the problem in this way (see Figure 6–3).

Jeremy and Dustin create and solve the problem in this way (see Figure 6–4).

Each group uses their interests to make the algebraic problem personally relevant. The students are motivated to create models to show how to represent the answer to the problem. In this way interests are utilized to differentiate the mathematics instruction.

Incorporating students' learning frameworks serves as a powerful mechanism for differentiating instruction in mathematics. Learning styles, multiple intelligences, environmental preferences, affective needs, and interests are ways to connect students with positive, prolific learning experiences. Students' motivation and commitment are increased as teachers incorporate learning frameworks.

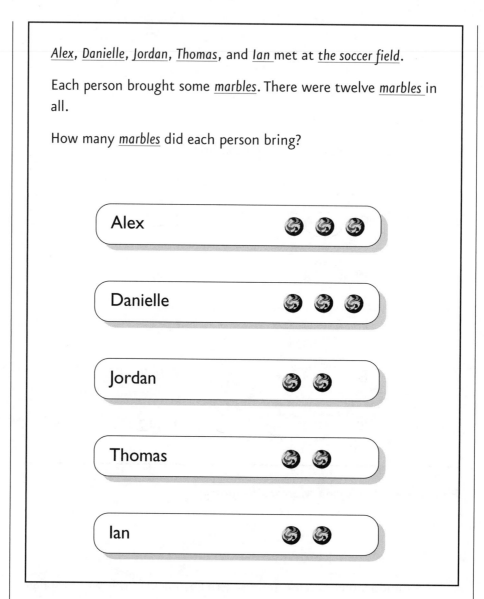

Alex, Danielle, Jordan, Thomas, and Ian met at the soccer field.

Each person brought some marbles. There were twelve marbles in all.

How many marbles did each person bring?

Figure 6–3 *Alex and Danielle's Response*

Jeremy, Dustin, Mike, Ryan, and Cory met at the park.

Each person brought some trading cards.

There were twelve trading cards in all.

How many trading cards did each person bring?

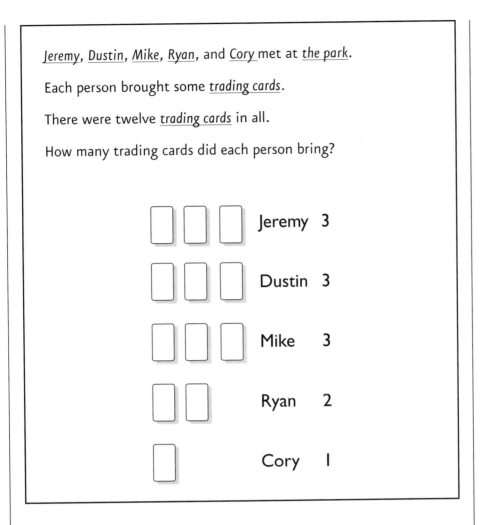

Jeremy 3

Dustin 3

Mike 3

Ryan 2

Cory 1

Figure 6–4 *Jeremy and Dustin's Response*

CHAPTER 7

Personal Assessment in Algebra

T-Charts Show Personal Assessments of Algebra

See:
Number, p. 148
Data, p. 87
Geometry, p. 90
Measurement, p. 90

One of the most basic graphic organizers is a simple yes/no chart. Using the yes/no labels in the form of a T-chart, students are asked if they are comfortable with a specific math situation. For example, before beginning a review of the commutative property with second graders, the teacher may give the problem $5 + 11 = 11 + \square$ and ask students if they know how to solve and explain it.

Each student places a name card in the *yes* or *no* column. Sometimes instead of name cards the class can use clothespins with individual students' names on each. Each clothespin is clipped on the *yes* or *no* side of the chart by the students. The teacher can take a quick look and see which students believe they need to review the topic and which students believe they need an additional challenge, as shown in Figure 7–1.

After working with her first-grade class on repeating patterns, the teacher asked the students to place their names on the yes/no chart. The question on the chart (see Figure 7–2) was *Can you show a repeating pattern using only shapes?*

Most of the children felt comfortable with modeling repeating patterns using only shapes. The teacher met with Jamal, Zoe, and Ian in a small group setting. She first complimented the students on their courage and honesty. Then she targeted the instruction to meet their academic needs. By the end of

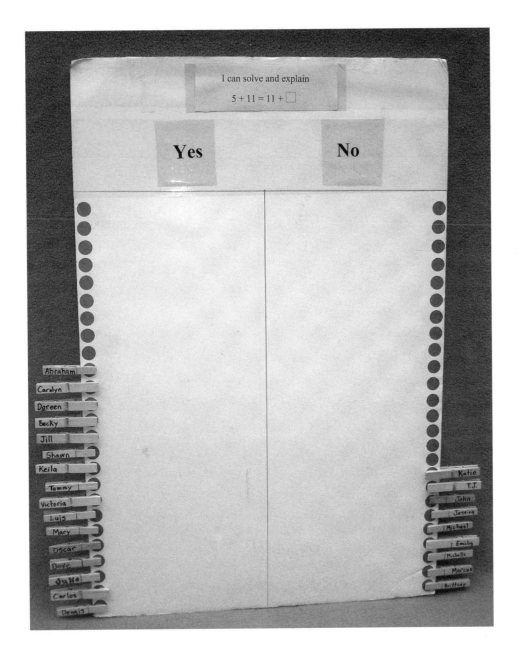

Figure 7–1 *A T-Chart Used in a Class to Show Students' Knowledge*

the small group session, the three students felt comfortable enough to move their names to the *yes* side of the chart.

Using graphic organizers offers a twofold benefit. The teacher gathers important personal assessment data and students have more opportunities to engage in mathematics because the data can be used instructionally. Many times the teacher talks with the students about what the data show. Using math vocabulary as they analyze and interpret the information revealed in the graph offers children additional opportunities to experience real-world mathematics.

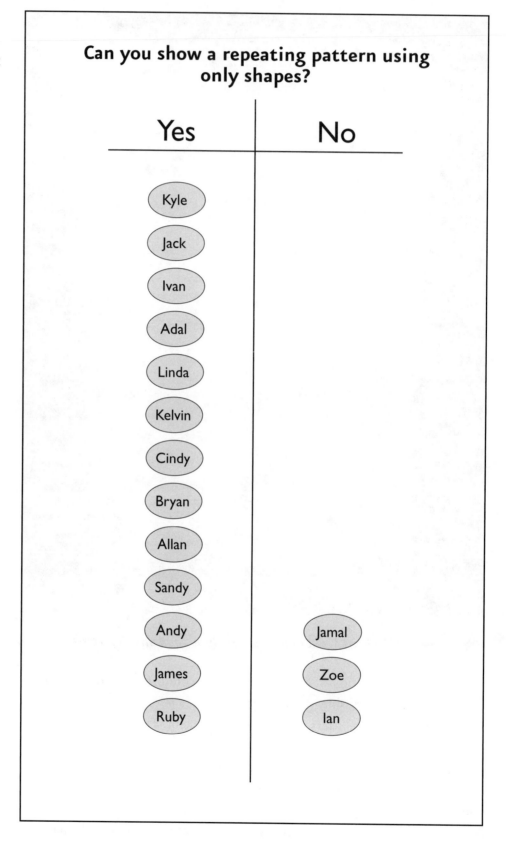

Figure 7–2 *An Example of a Yes/No Chart*

Bar Graphs Show Personal Assessments of Algebra

Bar graphs are also powerful tools. For example, before beginning a lesson on growing patterns, the teacher may put up a bar graph shell titled *How much I know about Growing Patterns*. The labels include a large balloon (I know a lot), a medium-size balloon (I know some), and a small balloon (I know a little bit). Students use sticky note cards to place their names in the category that represents how much they know (see Figure 7–3).

See:
Number, p. 150
Data, p. 89
Geometry, p. 91
Measurement, p. 93

The information is used to form groups and target instruction. After the lesson, students can change the placement of their name cards if they have revised their comfort levels with the topic. They may write in their math journals about how their level of understanding of a given topic has changed.

Comparison Circles Show Personal Assessments of Algebra

Comparison circles encourage math thinking and give excellent information. The teacher may use a single hoop or two (or more!) intersecting hoops. During a lesson on composing and decomposing numbers, the teacher asked the students to place their names inside the hoop or outside of the hoop to indicate

See:
Number, p. 152
Data, p. 89
Geometry, p. 91
Measurement, p. 93

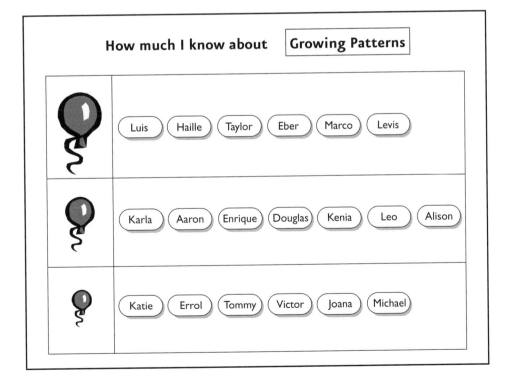

Figure 7–3
An Example of a Bar Graph

their current knowledge. The title of the hoop, shown in Figure 7–4, was *I know how to show 15 with base-ten blocks.*

Using the data presented by the students, the teacher offered a two-tiered lesson. One group worked on showing larger numbers in more than one way. The other group worked on showing numbers under twenty in at least one way. Everyone worked on composing and decomposing numbers and the tasks were suited to meet the academic needs of the students as they perceived them.

The teacher used two intersecting hoops as the graphic organizer and asked the students to tell what they know about showing equality. The title of the comparison circle graphic in Figure 7–5 was *I know how to show equality.* One hoop was labeled *with objects* and the other hoop was labeled *with numbers.*

The data indicate that many of the students are comfortable with showing equality using objects, but some of the students are not yet comfortable with showing equality with either objects or numbers. Several students are comfortable showing equality with numbers *and* objects. Only one student is comfortable showing equality with numbers but not with objects. The teacher

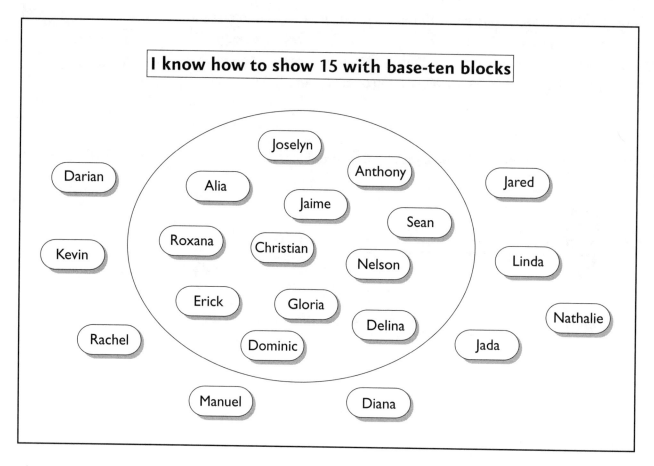

Figure 7–4 *Using a Hoop to Show Students' Current Knowledge*

can offer differentiated instruction to the students using the data presented in the comparison circles. The teacher has many options for how to differentiate the instruction. He can meet with each of the small groups and focus on what the students do not yet know. For example, the group that knows how to show equality with objects may work on showing equality with numbers. With the group that already knows how to show equality with both objects and numbers, the teacher could focus on showing equality with more challenging equations or showing equality using complex symbols.

Another option for the teacher is to give students a task they can work on independently or with each other while the teacher meets with small groups. If the students are working without the teacher, it may be most appropriate for the students to expand what they already know rather than move into something they are not yet comfortable with. For example, the students who know how to show equality with objects could create balanced situations with objects for each other to solve or describe. In this way, the students are working on something new, but they have some comfort with the topic. Therefore they can temporarily engage in meaningful tasks without direct instruction from the teacher.

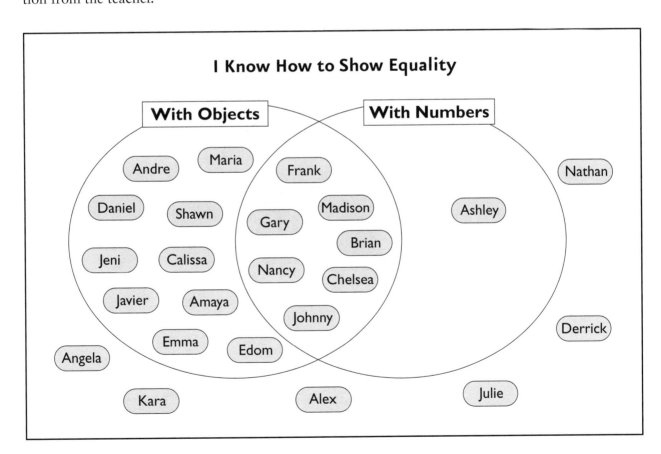

Figure 7–5 *Comparison Circles Show Students' Current Knowledge*

See:
Number, p. 154
Data, p. 91
Geometry, p. 96
Measurement, p. 96

Pyramids Show Personal Assessments of Algebra

Using the pyramid as the graphic organizer, the teacher asked the students to place name cards in ways that show level of understanding. The title of the pyramid, shown in Figure 7–6, was *I understand why this equation is correct: 3 + 2 + 1 = 4 + 2.*

The pyramid graphic allows students to show even more than one of three levels. They can actually place their names to designate levels within levels, such as high-middle or low-high. Some of the students strategically placed their name cards on the lines dividing the sections. Lenox shared, "I am between low and middle." Kathy said, "I am almost all the way in the high part." These

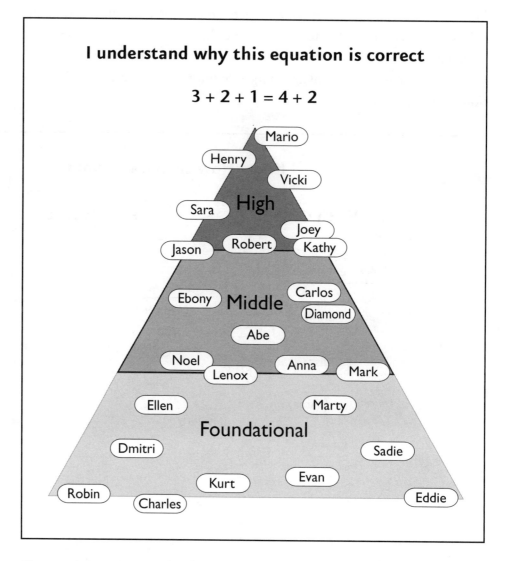

Figure 7–6 *An Example of a Pyramid*

comments and others indicate that the students are quite able to articulate their level of understanding. It is empowering to know where you are in the learning journey. The teacher can use the data to differentiate mathematics instruction.

Line Plots Show Personal Assessments of Algebra

Using a complex line plot that ranged from zero to one, the teacher asked the students to indicate their level of knowledge. The title of the line plot was *How much I know about decreasing patterns*. While the students had many experiences working with number lines and values between zero and one, this was one of the first personal assessment opportunities given to this class so the teacher decided to make the pieces of data anonymous. The teacher hung the line plot behind the easel and allowed each student to privately place an X on the line (see Figure 7–7). Sometimes students need several opportunities to use a graphic organizer in an anonymous fashion before they are ready to publicly display their levels of knowledge or comfort.

See:
Number, p. 155
Data, p. 93
Geometry, p. 97
Measurement, p. 96

The data show three distinct groups. The teacher can use the data to offer three different levels of the same task and allow students to choose the level that matches their comfort. Or the teacher can help students form heterogeneous groups to work on a given task. Or the teacher can offer small group work at various levels and allow the children to choose which group to attend. Or perhaps the teacher uses an informal assessment to differentiate the instruction and the children use a different color X to privately indicate their

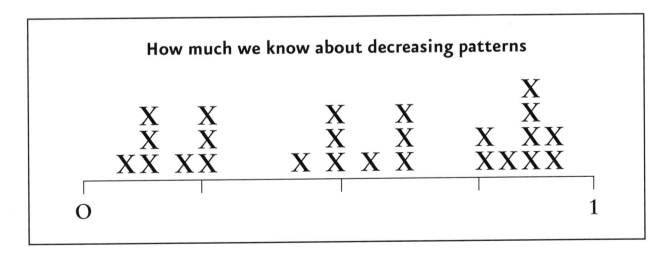

Figure 7–7 *A Line Plot*

new level of knowledge following the instruction. Afterward, the class can analyze and interpret the data, highlighting growth trends.

Fractions Grids Show Personal Assessments of Algebra

See:
Number, p. 157
Data, p. 95
Geometry, p. 98
Measurement, p. 98

After a lesson on representing quantitative relationships, the teacher posed this fractions grid and title: *I can show this math situation: Twelve cookies. Three cookies on each plate.* Students who were completely confident with showing four groups of three cookies with words, numbers, or pictures placed their name stars in the 1 whole section of the graphic. Students with solid knowledge (but not complete confidence) of how to do so placed their name stars in the ¾ section of the graphic. Students with some knowledge of how to do so placed their name stars in the ½ section of the graphic. Students with a little bit of knowledge of how to do so placed their name stars in the ¼ section of the graphic. If any students had zero knowledge of how to do so, they could place their name stars outside of the graphic (see Figure 7–8).

The teacher used the data to plan for the next day's instruction. She offered a challenge for the students who indicated that they completely understood how to represent the math situation. She offered targeted instruction to the ¾ group in a way that helped them move to complete understanding. She built more foundations with the ½ group and the ¼ group by targeting instruction. All of the students increased their levels of understanding.

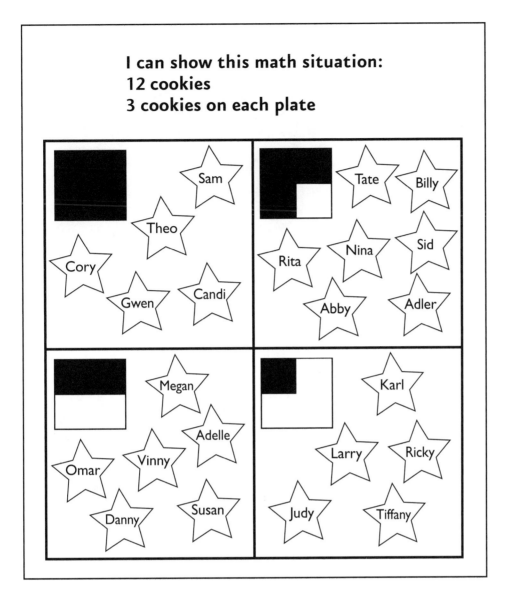

I can show this math situation:
12 cookies
3 cookies on each plate

Figure 7-8 *An Example of a Fraction Grid*

REFERENCES AND ADDITIONAL READING

Anderson, L., and Krathwohl, eds. 2001. *A Taxonomy for Learning, Teaching, and Assessing: A Revision of Bloom's Taxonomy of Educational Objectives.* New York: Longman.

Barnett-Clarke, C., and A. Ramirez. 2004. "Language Pitfalls and Pathways to Mathematics." In *Perspectives on the Teaching of Mathematics*, edited by R. Rubenstein. Reston, VA: National Council of Teachers of Mathematics.

Behr, M., S. Erlwanger, and E. Nichols. 1980. "How Children View the Equal Sign." *Mathematics Teaching* 92: 13–15.

Bloom, B., M. Englehart, E. Furst, W. Hill, and D. Krathwohl. 1956. *Taxonomy of Educational Objectives: The Classification of Educational Goals—Handbook I: Cognitive Domain.* New York: Longmans Green.

Brewer, D., D. Rees, and L. Argys. 1995. "Detracking America's Schools: The Reform Without Costs?" *Phi Delta Kappa* 77 (3): 210–215.

Caine, R., and G. Caine. 1991. *Making Connections: Teaching and the Human Brain.* Alexandria, VA: Association for Supervision and Curriculum Development.

Checkley, K. 2001. "Algebra and Activism: Removing the Shackles of Low Expectations—A Conversation with Robert P. Moses." *Educational Leadership* 59 (2): 6–11.

Clements, D., and J. Sarama, eds. 2004. *Engaging Young Children in Mathematics: Standards for Early Childhood Mathematics Education.* Mahwah, NJ: Lawrence Erlbaum Associates.

Copley, J. 2000. *The Young Child and Mathematics.* Washington D.C.: National Association for the Education of Young Children.

Dunn, R., K. Dunn, and G. Price. 1985. *Learning Style Inventory.* Lawrence, KS: Price Systems.

Dunn, R., and S. A. Griggs, eds. 2000. *Practical Approaches to Using Learning Styles in Higher Education.* Westport, CT: Bergin & Garvey.

Falkner, K., L. Levi, and T. Carpenter. 1999. "Children's Understanding of Equality: A Foundation for Algebra." *Teaching Children Mathematics* 6 (4): 232–236.

Gallagher, J. 1993. "Ability Grouping: A Tool for Educational Success." *College Board Review* 168 (Summer): 21–27.

Gardner, H. 1983. *Frames of Mind*. New York: Basic Books.

Gardner, H. 1993. *Multiple Intelligences: The Theory in Practice*. New York: Basic Books.

Gardner, H. 1999. *Intelligence Reframed: Multiple Intelligences for the 21st Century*. New York: Basic Books.

Gentry, M., and S. Owen. 1999. "An Investigation of the Effects of Total School Flexible Cluster Grouping on Identification, Achievement, and Classroom Practices." *Gifted Child Quarterly* 43 (4): 224–242.

Goldring, E. 1990. "Assessing the Status of Information on Classroom Organizational Frameworks for Gifted Students." *Journal of Educational Research* 83 (6): 313–326.

Gordon, T. 1974. *T.E.T.: Teacher Effectiveness Training*. New York: Peter H. Wyden.

Hart, L. 1981. "Do Not Teach Them, Help Them Learn." *Learning* 9 (8): 39–40.

Jarvis, P. 1992. "Reflective Practice and Nursing." *Nurse Education Today* 12 (3): 174–181.

Jensen, E. 1998. *Teaching with the Brain in Mind*. Alexandria, VA: Association for Supervision and Curriculum Development.

Kagan, S. 1997. *Cooperative Learning*. San Clemente, CA: Resources for Teachers.

Kramer, M. J. 2005. "The Key to Raising Achievement: Four Guiding Principles." *AASA New Superintendents E-Journal* (3). http://www.aasa.org/publications/content.cfm?ItemNumber=7039, April 25, 2008.

Kulik, J. A. 1992. *An Analysis of the Research on Ability Grouping: Historical and Contemporary Perspectives*. Storrs, CT: National Research Center on the Gifted and Talented.

Kulik, J. A., and C. C. Kulik. 1992. "Meta-Analytic Findings on Grouping Programs." *Gifted Child Quarterly* 36 (2): 73–77.

Lott, Johnny W., ed. 2000. "Algebra? A Gate? A Barrier? A Mystery!" *Mathematics Education Dialogues* 3 (2): 1–12.

Lou, Y., P. C. Abrami, J. C. Spence, C. Poulsen, B. Chambers, and S. d'Apollonia. 1996. "Within-Class Grouping: A Meta-Analysis." *Review of Educational Research* 66 (4): 423–458.

Loveless, T. 1998. *The Tracking and Ability Grouping Debate*. Washington, D.C.: Thomas B. Fordham Foundation.

Loveless, T. 1999. "Will Tracking Reform Promote Social Equity?" *Educational Leadership* 56 (7): 28–32.

Molina, M., and C. Ambrose. 2006. "Fostering Relational Thinking While Negotiating the Meaning of the Equals Sign." *Teaching Children Mathematics* 13 (2): 111–117.

National Association for the Education of Young Children and National Council of Teachers of Mathematics. 2002. *Early Childhood Mathematics: Promoting Good Beginnings, A Joint Position Statement*. Reston, VA: National Council of Teachers of Mathematics.

National Council of Teachers of Mathematics. 2000. *Principles and Standards for School Mathematics*. Reston, VA: National Council of Teachers of Mathematics.

Oakes, J. 1985. *Keeping Track: How Schools Structure Inequality.* New Haven, CT: Yale University Press.

Oakes, J. 1988. "Tracking in Mathematics and Science Education: A Structural Contribution to Unequal Schooling." In *Class, Race, and Gender in American Education,* edited by L. Weiss. Albany, NY: State University of New York Press.

Oakes, J. 1990. *Multiplying Inequalities: The Effects of Race, Social Class, and Tracking on Opportunities to Learn Mathematics and Sciences.* Santa Monica, CA: Rand Corporation.

Ogle, D. S. 1986. "K-W-L Group Instructional Strategy." In *Teaching Reading as Thinking,* edited by A. S. Palincsar, D. S. Ogle, B. F. Jones, and E. G. Carr. Alexandria, VA: Association for Supervision and Curriculum Development.

Olszewski-Kubilius, P. 2003. "Is Your School Using Best Practices of Instruction for Gifted Students?" *Talent Newsletter.* Evanston, IL: Center for Talent Development, Northwestern University. http://www.ctd.northwestern.edu/resources/talentdevelopment/bestinstruction.html, May 5, 2008.

Reis, S. M., D. E. Burns, and J. S. Renzulli. 1992. *Curriculum Compacting: The Complete Guide to Modifying the Regular Curriculum for High Ability Students.* Mansfield Center, CT: Creative Learning Press.

Renninger, K., and S. Hidi. 1992. *The Role of Interest in Learning and Development.* Mahwah, NJ: Lawrence Erlbaum Associates.

Renzulli, J. S., and L. H. Smith. 1978. *The Compactor.* Mansfield Center, CT: Creative Learning Press.

Renzulli, J. S., L. H. Smith, and S. M. Reis. 1982. "Curriculum Compacting: An Essential Strategy for Working with Gifted Students." *The Elementary School Journal* 82: 185–194.

Richardson, K. 2004. "Making Sense." In *Engaging Young Children in Mathematics: Standards for Early Childhood Mathematics Education,* edited by D. Clements and J. Sarama. Mahwah, NJ: Lawrence Erlbaum Associates.

Senz-Ludlow, A., and C. Walgamuth. 1998. "Third Graders' Interpretations of Equality and the Equal Symbol." *Educational Studies in Mathematics* 35: 153–187.

Slavin, R. E. 1987a. "Ability Grouping and Its Alternatives: Must We Track?" *American Educator* 11 (2): 32–36, 47–48.

Slavin, R. E. 1987b. "Ability Grouping and Student Achievement in Elementary Schools: A Best Evidence Synthesis." *Review of Educational Research* 57 (3): 293–336.

Slavin, R., H. Braddock, and H. Jomills. 1993. "Ability Grouping: On the Wrong Track." *College Board Review* 168: 11–18.

Smith-Maddox, R., and A. Wheelock. 1995. "Untracking and Students' Futures: Closing the Gap Between Aspirations and Expectations." *Phi Delta Kappa* 77 (3): 222–228.

Taylor-Cox, J. 2003. "Algebra in the Early Years? Yes!" *Young Children* 58 (1): 14–21.

Tomlinson, C. 1995. *How to Differentiate Instruction in Mixed-Ability Classrooms.* Alexandria, VA: Association for Supervision and Curriculum Development.

Tomlinson, C. 1996. *How to Differentiate Instruction in Mixed-Ability Classrooms Professional Inquiry Kit.* Alexandria, VA: Association for Supervision and Curriculum Development.

Tomlinson, C. 1999. *The Differentiated Classroom: Responding to the Needs of All Learners*. Alexandria, VA: Association for Supervision and Curriculum Development.

Vygotsky, L. 1962. *Thought and Language*. Cambridge, MA: MIT Press.

Wheelock, A. 1992. *Crossing the Tracks: How Untracking Can Save America's Schools*. New York: New Press.

Wiggins, G. 2003. "'Get Real!' Assessing for Quantitative Literacy." In *Quantitative Literacy: Why Numeracy Matters for Schools and Colleges*, edited by B. Madison and L. Arthur. Princeton, NJ: National Council on Education and the Disciplines.

Wiggins, G., and J. McTighe. 1998. *Understanding by Design*. Alexandria, VA: Association for Supervision and Curriculum Development..

Wood, D., J. Bruner, and G. Ross. 1976. "The Role of Tutoring in Problem-Solving." *Journal of Child Psychology and Psychiatry* 17 (2): 89–100.

INDEX